Table of Contents

INTRODUCTION 1

I. OVERVIEW AND NEED FOR FRAMEWORK 1

    A.   Role of Information and Communication Technology in Business and Society 1

    B.   Accessible Technology—The Gateway Civil Rights Issue for People with Disabilities 1

    C.   Legal/Policy Context 2

    D.   The Business Case and Emerging Business Practices 3

    E.   Values and Guiding Principles—the Concept of Universal Design 3

    F.   Building the Framework for an Accessible ICT Strategic Plan 4

II. ACTION STEP 1: UNDERSTAND THE TERMINOLOGY AND TARGET POPULATION .. 4

    A.   The Terminology 5

    B.   The Target Population 8

III. ACTION STEP 2: UNDERSTAND THE POLICY FRAMEWORK 9

    A.   Federal Civil Rights Laws 10

        The Americans with Disabilities Act (ADA) 10

        Sections 501, 503 and 504 of the Rehabilitation Act of 1973, as amended 11

        Applicability of ADA and Sections 501, 503 and 504 of the Rehabilitation Act to Accessible ICT, including the Internet 11

    B.   Federal Standards and International Guidelines Regarding Accessible ICT 12

        Section 508 of the Rehabilitation Act of 1973, as amended 12

        Section 255 of the Telecommunications Act of 1996 13

    C.   International Accessibility Standards Applicable to Websites 13

    D.   Applicability of Section 508 and WCAG 2.0 to Private Sector Employers 15

    E.   United Nations Convention on the Rights of Persons with Disabilities 16

    F.   Summary—Importance of Understanding the Policy Framework 16

IV. ACTION STEP 3: UNDERSTAND THE BUSINESS CASE 18

V. ACTION STEP 4: SELF-ASSESS ICT PRACTICES, SET GOALS AND ESTABLISH PRIORITIES FOR IMPROVEMENT 20

VI. ACTION STEP 5: ADVANCE CORPORATE POLICIES, PRACTICES AND PROCEDURES 21

    A.   Introduction 21

    B.   Software Applications and Operating Systems 21

    C.   Websites 21

    D.   Telecommunications Products 23

    E.   Video or Multimedia Products 24

F.   Self-Contained, Closed Products 24

G.   Desktop and Portable Computers..24

H.   Information, Documentation and Support 25

I.   Online Application Systems 25

VII. ACTION STEP 6: ESTABLISH AN INFRASTRUCURE FACILITATING IMPLEMENTATION OF THE POLICIES **28**

A.   Leadership and Team Approach (Managers Across Divisions) 28

B.   Outsourcing and Procurement...29

    1.   Outsourcing...29

    2.   The Procurement Process..29

    3.   Role of Procurement or Contracting Officials 31

C.   Accessibility Training for In-House Staff..32

D.   Deployment..33

VIII. ACTION STEP 7: COMMIT TO ACCOUNTABILITY AND CONTINUOUS IMPROVEMENT **33**

A.   Appointment of ADA Coordinator, Chief Technology Accessibility Officer and Designation of Accountable Managers..33

B.   Notice of Policy and Responsible Persons and Offices 35

C.   Participation by Individuals with Disabilities and Outside Experts............35

D.   Measurable Objectives, Benchmarks and Prioritizing Areas Needing Improvement....................35

    Prioritizing by Barriers 36

    Prioritizing by Area...37

E.   Monitoring, Auditing, Reporting and Continuous Improvement................38

IX. ACTION STEP 8: IDENTIFY AVAILABLE RESOURCES................................**39**

A.   General Regulatory Guidance (on Sections 503, 504 and 508 of the Rehabilitation Act of 1973, as amended)..39

B.   Understanding the Business Case 39

C.   Acquiring and Purchasing Accessible Technology 39

D.   Understanding Assistive Technologies..40

    Vision Impairments..40

    Dexterity and Mobility Impairments 40

    Hearing Impairments 40

    Learning Impairments...40

    Language and Communication Impairments 40

E.   Accessibility Testing Tools...41

F.   Accessibility Training Resources..41

G.   Implementing Accessible Software Applications and Operating Systems...................41

H.   Creating Accessible Websites...42

I.   Creating Accessible Online Job Applications...42

J.   Creating Accessible Documents 42

K.   Selecting Accessible Telecommunications Products (mobile phones, etc.) 42

L.   Creating Accessible Videos and Multi-media Products..43

M.   Accessibility of Desktop and Portable Computers 43

N.   Accessibility of Social Media 43

O.   Sample Accessibility Evaluation Tools and Checklists.......................................43

P.   Best Practices...44

X.   APPENDIX: COMPREHENSIVE BENCHMARKING TOOL............................45

A.   ACTION STEP 1: UNDERSTAND THE TERMINOLOGY AND TARGET POPULATION...45

B.   ACTION STEP 2: UNDERSTAND THE POLICY FRAMEWORK 46

C.   ACTION STEP 3: UNDERSTAND THE BUSINESS CASE 47

D.   ACTION STEP 4: SELF-ASSESS ICT PRACTICES, SET GOALS AND ESTABLISH PRIORITIES FOR IMPROVEMENT 47

E.   ACTION STEP 5: ADVANCE CORPORATE POLICIES, PRACTICES AND PROCEDURES…..48

F.   ACTION STEP 6: ESTABLISH AN INFRASTRUCURE FACILITATING IMPLEMENTATION OF THE POLICIES 56

G.   ACTION STEP 7: COMMIT TO ACCOUNTABILITY AND CONTINUOUS IMPROVEMENT. 60

H.   ACTION STEP 8: IDENTIFY AVAILABLE RESOURCES......................................64

XI. ENDNOTES..65

INTRODUCTION:

As technology continues to transform the workplace, demand is growing for the development, purchase, maintenance and use of information and communication technology (ICT)[1] that is accessible to and usable by all applicants and employees, including individuals with disabilities. Leading companies recognize that fostering an accessible workplace is the smart thing to do, both from a business standpoint and a legal perspective. This paper provides a framework that can be used to develop technical assistance tools to help employers (including government contractors) design, purchase, lease, maintain and use ICT that is accessible to and usable by people with disabilities and others.

More specifically, the Framework can be used as a tool for self-assessment to make easy point-by-point comparisons with current strategies and practices. In addition, the Framework can help a company develop a corporate-wide, comprehensive strategic plan (affirmative action program, where applicable) for adopting and implementing accessible ICT policies, practices and procedures for assessing progress made over time and ensuring continuous improvement. Further, the Framework can be used as a tool for internal staff development and training.

In addition to the Framework, the paper also includes a comprehensive Benchmarking Tool, located in the Appendix, which companies could use to conduct a self-assessment and create their own corporate-wide, comprehensive strategic plan (affirmative action program, where applicable), including accountability mechanisms and methods for ensuring continuous improvement. This Benchmarking Tool is organized in accordance with the key components of an Accessible ICT Strategic Plan outlined herein.

I. OVERVIEW AND NEED FOR FRAMEWORK

A. Role of Information and Communication Technology in Business and Society

Information and communication technologies (ICT), particularly Web-based Internet and Intranet information and applications, play a significant and ever-expanding role in everyday life in the United States. The role of ICT impacts commerce through the purchase of goods and services and how we exchange information. In addition, ICT has become a major gateway to employment, primarily because recruiting and hiring systems are often Web based. In many cases, the only way to apply for a job or to sign up for an interview is via the Internet. Additionally, job applicants research employment opportunities online and use the Internet to learn about potential employers' needs and policies, as well as inquire about job opportunities. Furthermore, companies are using Intranet Web sites to conduct job-related testing, provide training to their employees, and share information about fringe benefits and corporate events and activities.

B. Accessible Technology—The Gateway Civil Rights Issue for People with Disabilities

In light of the critical role ICT plays in contemporary society, including in the employment context, the federal government is now recognizing accessible ICT as the most critical civil rights issue for people with disabilities. The Department of Justice (DOJ) recently recognized that:

> "As more and more of our social infrastructure is made available on the Internet – in some cases, exclusively online – access to information and electronic technologies is increasingly becoming the gateway civil rights issue for individuals with disabilities."[2]

Faced with the reality that ICT can open doors for people with disabilities, the federal government is also recognizing that technological advances will leave people with disabilities behind if employers do not purchase and use accessible ICT and ICT developers and manufacturers do not make their new products accessible. In carrying out its responsibilities under the Americans with Disabilities Act (ADA) and the Rehabilitation Act of 1973, as amended, including Section 501 (affirmative action for federal agencies), Section 503 (affirmative action by government contractors), Section 504 (nondiscrimination by recipients of federal aid), and Section 508 (procurement of electronic and information technology by federal agencies), the federal government is now recognizing that it:

> "must make sure that the legal protections for the rights of individuals with disabilities are clear and sufficiently strong to ensure that innovation increases opportunities for everyone. We must avoid the travesty that would occur if the doors that are opening to Americans from advancing technologies were closed for individuals with disabilities because we were not vigilant."[3]

C. Legal/Policy Context

DOJ has made the following statements regarding the need to provide greater clarity in official policy pronouncements regarding accessible ICT, particularly access to the Internet. According to DOJ:[4]

- "Voluntary standards have generally proved to be sufficient where obvious business incentives align with discretionary governing standards as, for example, with respect to privacy and security standards designed to increase consumer confidence in e-commerce. There has not, however, been equal success in the area of accessibility."

- "The [Web Accessibility Initiative of the World Wide Web Consortium] (WAI) leadership has recognized this challenge and has stated that in order to improve and accelerate Web accessibility it is important to "communicat[e] the applicability of the Americans with Disabilities Act to the Web more clearly, with updated guidance * * * ." *Achieving the Promise of the Americans with Disabilities Act in the Digital Age – Current Issues, Challenges, and Opportunities: Hearing Before the Subcommittee on the Constitution, Civil Rights and Civil Liberties, H. Committee On the Judiciary*, 111th Cong. (Apr. 22, 2010) (statement of Judy Brewer, Director, Web Accessibility Initiative at the W3C)."

- "It is clear that the system of voluntary compliance has proved inadequate in providing website accessibility to individuals with disabilities. *See, e.g.,* National Council on

Disability, *The Need for Federal Legislation and Regulation Prohibiting Telecommunications and Information Services Discrimination* (Dec. 19, 2006), available at http://www.ncd.gov/publications/2006/Dec282006 (last visited June 29, 2010) (discussing how competitive market forces have not proven sufficient to provide individual with disabilities access to telecommunications and information services)."

Thus, it appears that the federal government is recognizing the need to take a more active role in ensuring, through the formal regulatory process, that people with disabilities have effective and meaningful access to ICT under our nation's civil rights laws, including Internet-based information and applications.

D. The Business Case and Emerging Business Practices

In addition to recognizing the evolving legal and policy context in the United States and around the world, many companies are choosing to design, purchase and deploy accessible ICT for other business reasons. The business case for accessible ICT includes improved efficiency by addressing the personal needs of all ICT users, including those with sensory, physical and mental impairments, age-related limitations and temporary limitations, as well as those who are novices. Accessible technology also helps businesses support workforce diversity, recruit from a larger pool of candidates, and enhance team collaboration and communication among all employees, including those with disabilities.[5]

Best and promising business practices regarding accessible ICT are emerging. These business practices include securing support from executive leadership, such as the Chief Executive Officer (CEO), and a company-wide involvement and designation of respective roles and responsibilities by key managers (team approach), including the Chief Information Officer, the Chief Acquisition Officer, the Chief Procurement Officer, human resources, education and training, financial and marketing. These business practices also include:

- The development and implementation of a comprehensive strategic action plan that includes assessing needs and current levels of accessibility of websites and other ICT used in the company;
- Setting accessibility goals and establishing priorities;
- Adopting policies, practices and procedures regarding design, content production, purchasing, training and deployment;
- Establishing a corporate-wide infrastructure; and
- Adopting monitoring, accountability and continuous improvement mechanisms and strategies to meet the goals, including the establishment of measurable objectives and benchmarks and the appointment of a Chief Technology Accessibility Officer.[6]

E. Values and Guiding Principles—the Concept of Universal Design

Many of the companies that are adopting accessible ICT policies, practices and procedures are doing so in part based on recognition of the concept/philosophy of "universal design" (i.e., designing, purchasing, leasing, maintaining and deploying products and services that are usable by people with the widest possible range of functional capabilities). This includes products and services that are directly usable (without requiring assistive technologies) and those that are made compatible or interoperable with assistive technologies.[7]

Accessible ICT, including accessible websites, offer significant advantages that go beyond access for people with disabilities. Some of these benefits include reducing website maintenance time, reducing bandwidth requirements and server load, enabling content repurposing on different devices, and simplifying forward migration to advanced standards that provide expanded capabilities for business websites.[8]

F. Building the Framework for an Accessible ICT Strategic Plan

As stated in the introduction, this paper provides a framework that can be used to develop technical assistance tools to help employers (including government contractors) design, purchase, lease, maintain and use ICT that is accessible to and usable by people with disabilities and others. More specifically, this paper serves as a framework for developing and implementing an accessible ICT strategic plan that includes the following core action steps:

- Understand Key Terminology and Target Population
- Understand the Policy Framework
- Understand the Business Case
- Self-Assess ICT Practices, Set Goals and Establish Priorities for Improvement
- Advance Corporate Policies, Practices and Procedures
- Establish an Infrastructure to Facilitate Implementation of the Policies
- Commit to Accountability and Continuous Improvement
- Identify Available Resources

The Framework may be used as a tool for self-assessment to make easy point-by-point comparisons with current strategies and practices. The Framework can also be used to help guide a company's thinking about the best course for developing a corporate-wide, comprehensive strategic plan (affirmative action program, where applicable) for adopting and implementing accessible ICT policies, practices and procedures, for assessing progress made over time, and ensuring continuous improvement. Further, the Framework can be used as a tool for internal staff development and training.

This Framework does not create any new legal requirements or change current legal requirements. Instead, it lists examples of strategies and practices that employers have found to be effective in enhancing employment opportunities for qualified individuals with disabilities through accessible ICT.

II. ACTION STEP 1: UNDERSTAND THE TERMINOLOGY AND TARGET POPULATION

The first action step a company may want to take in its efforts to meet the needs of qualified individuals with disabilities through the development, procurement, maintenance and use of accessible ICT is to review the key terms used to describe accessible ICT and the target populations affected by inaccessible ICT systems. Frequently used acronyms are also identified and their meaning articulated.

A. The Terminology[9]

Access Board. The Access Board is an independent federal agency that develops and maintains, among other things, accessibility guidelines and standards, provides technical assistance and training on the guidelines and standards, and enforces accessibility standards for federally funded facilities. Section 508 of the Rehabilitation Act requires the Access Board to publish standards setting forth a definition of electronic and information technology (also referred to as information and communication technology, ICT) and technical and performance standards. The Access Board's official name is the Architectural and Transportation Barriers Compliance Board.

Accessible technology. Technology that can be used by people with a wide range of abilities and disabilities. It incorporates the principles of universal design. Each user is able to interact with the technology in ways that work best for him or her. Accessible technology is either directly accessible—in other words, it is usable without assistive technology—or it is compatible with standard assistive technology.[10]

Alternate formats. Alternate formats usable by people with disabilities may include, but are not limited to: Braille, ASCII text, large print, Web, e-mail, CD-ROM/DVD, recorded audio, other electronic formats.

Alternative methods. Alternate methods are different means of providing information, including product documentation, to people with disabilities. Alternate methods may include, but are not limited to: voice, fax, relay service, TTY, Web posting, e-mail, sign language, instant messaging, mobile texting, captioning, audio description, text to speech syntheses.

ADA. The Americans with Disabilities Act of 1990 is a comprehensive civil rights law prohibiting discrimination on the basis of disability in the areas of employment, public services, public accommodations and telecommunication relay services for hearing impaired and speech impaired individuals.

Assistive technology. Any item, piece of equipment or system, whether acquired commercially, modified or customized, that is commonly used to increase, maintain or improve functional capacities of individuals with disabilities. The term includes traditional assistive technology hardware and software along with mainstream technology used for assistive purposes, virtual assistive technology delivered as a Web service and integration of products into a system that provides assistive technology functions that allow individuals with disabilities to access information and communication technology. Examples include:

- Screen enlargers that act like magnifiers to help people with low vision;
- Onscreen keyboards that allow people who are unable to use a standard keyboard to select keys using methods such as a pointing device or switch;
- Voice recognition (instead of using a mouse or keyboard);

- Alternative input devices that enable individuals to control their computers through means other than a standard keyboard or pointing devices, e.g., head-operated pointing devices and sip and puff systems controlled by breathing; screen readers that allow users who are blind to hear what is happening on their computer by converting the screen display to digitized speech, etc.

Disability. As defined by the Americans with Disabilities Act, the term 'disability' means, with respect to an individual:

(A) a physical or mental impairment that substantially limits one or more major life activities of such individual;

(B) a record of such an impairment; or

(C) being regarded as having such an impairment.[11]

GPAT. Government Product/Service Accessibility Template is a solicitation documentation tool produced by the BuyAccessible Wizard to assist federal contracting and procurement officials in fulfilling the market research requirements associated with Section 508 regulations.

Information and communication technology (ICT), which is also referred to as electronic and information technology (EIT). ICT includes information technology and any equipment or interconnected system or subsystem of equipment, that is used in the creation, conversion or duplication of data or information. ICT also includes information technology and any equipment or interconnected system or subsystem of equipment, which is used in the automatic acquisition, storage, analysis, evaluation, manipulation, management, movement, control, display, switching, interchange, transmission, reception or broadcast of data or information. The term includes, but is not limited to, electronic content, including e-mail, electronic documents and Internet and Intranet websites; telecommunications products, including video communication terminals; computers and ancillary equipment, including external hard drives; software, including operating systems and applications; information kiosks and transaction machines; videos; IT services; and multifunction office machines that copy, scan and fax documents.

Note: "Electronic and information technology (EIT)" is a term used in the 1998 amendments to Section 508 of the Rehabilitation Act to define the scope of products covered under Section 508.[12]

Information technology. Any equipment or interconnected system or subsystem of equipment, that is used in automatic acquisition, storage, analysis, evaluation, manipulation, management, movement, control, display, switching, interchange, transmission or reception of data or information. The term information technology includes computers, ancillary equipment, including software, firmware and similar procedures, services (including support services) and related resources.

Interoperability. Assistive technology and ICT interoperability is the ability of assistive technology (both soft and hard) and standard ICT (both soft and hard) from multiple vendors to exchange and use information meaningfully and without adverse system consequences, or when possible, the need for special configuration or adaptation on the part of the user.

Online application system. An online application system includes, but is not limited to, all electronic or Web-based systems that an employer uses in all of its personnel activities.[13]

Operable controls. A component of a product that requires physical contact for normal operation. Operable controls include, but are not limited to, input and output trays, card slots, keyboards or keypads.

Platform accessibility services. Services provided by a platform enabling interoperability with assistive technology, such as but not limited to accessibility Application Programming Interfaces (API) or Document Object Model (DOM).

Section 501. Section 501 of the Rehabilitation Act of 1973, as amended prohibits discrimination on the basis of disability by federal agencies and requires such agencies to take affirmative action to hire and promote qualified individuals with disabilities.

Section 503. Section 503 of the Rehabilitation Act of 1973, as amended prohibits discrimination on the basis of disability by government contractors and requires such contractors to take affirmative action to hire and promote qualified individuals with disabilities.

Section 504. Section 504 of the Rehabilitation Act of 1973, as amended prohibits discrimination on the basis of disability by recipients of federal financial assistance.

Section 508. Section 508 of the Rehabilitation Act of 1973, as amended requires that when federal agencies develop, procure, maintain or use electronic and information technology, federal employees with disabilities have access to and use of information and data that is comparable to the access and use by federal employees who are not individuals with disabilities, unless an undue burden would be imposed on the agency. Section 508 also requires that individuals with disabilities who are members of the public seeking information or services from a federal agency have access to and use of information that is comparable to that provided to the public who are not individuals with disabilities, unless an undue burden would be imposed on the agency. Furthermore, it requires the Access Board to publish standards setting forth the definition of EIT, also referred to as ICT, and technical and functional performance standards for such technology.

Section 255 of the Telecommunications Act of 1966. Section 255 of the Telecommunications Act of 1996 requires manufacturers to ensure that telecommunications equipment or customer premises equipment are designed, developed and fabricated to be accessible to and usable by individuals with disabilities when it is readily achievable to do so and readily achievable means easily accomplishable without much difficulty or expense. The Access Board is given responsibility for developing accessibility guidelines for telecommunication equipment and customer premises equipment in conjunction with the Federal Communications Commission (FCC).

Telecommunications. The transmission, between or among points specified by the user, of information of the user's choosing, without change in the form or content of the information as sent and received.

TTY. An abbreviation for teletypewriter. Machinery or equipment that employs interactive text based communications through the transmission of coded signals across the telephone network. TTYs may include, for example, devices known as TDDs (telecommunication display devices or telecommunications devices for deaf persons) or computers with special modems. TTYs are also called text telephones.

Universal design. The concept or philosophy for designing and delivering products and services that is usable by people with the widest possible range of functional capabilities. This includes products and services that are directly usable (without requiring assistive technologies) and those that are made compatible with assistive technologies.

Video description. The insertion of verbal or auditory descriptions of on-screen visuals intended to describe important visual details that are contained in, or that cannot be understood from, the main audio output alone. Video descriptions supplement the regular audio track of a program and are usually inserted between dialogue narrations to provide information about actions, characters and on-screen text that appear without verbalization. Video descriptions are a way to let people who are blind or have low vision know what is happening on-screen.

VPAT. Voluntary Product Accessibility Template is a tool used to document a product's conformance with the accessibility standards under Section 508 of the Rehabilitation Act. The purpose of the VPAT is to assist federal contracting officials and other buyers in making preliminary assessments regarding the availability of commercial information and communication technology products and services with features that support accessibility.

WCAG. The Web Content Accessibility Guidelines are recognized voluntary international guidelines for Web accessibility created by the Web Accessibility Initiative (WAI) of the World Wide Web Consortium (W3C). These guidelines detail how to make Web content and Web applications accessible to individuals with disabilities. The most recent and updated version is WCAG 2.0, published in December 2008.

B. The Target Population

According to DOJ, millions of people have disabilities, including visual, auditory, physical, speech, cognitive and neurological disabilities, that affect their use of ICT.[14]

People who are **blind or have low vision** are often the most affected by inaccessible information and electronic technology. Many individuals with visual impairments use an assistive technology known as a screen reader that enables them to access the information on computers or websites. Screen readers read text aloud as it appears on the computer screen. Individuals who are blind may also use refreshable Braille displays, which convert the text of websites to Braille. Sometimes, those individuals will use keyboards in lieu of a mouse to move up and down on a screen or sort through a list and select an item.

People who have difficulty using a computer mouse because of **mobility impairments**, for example, may use an assistive technology that allows them to control software with verbal commands. But websites and other technologies are not always compatible with those assistive technologies. Challenges for persons with mobility impairments include navigating and interacting with Web pages (e.g., difficulty moving the cursor with the required precision) or lacking the requisite manual dexterity or hand-eye coordination to use a standard keyboard. They will often use mouse and keyboard alternative assistive devices or helper applications such as a head wand or voice recognition software.

Captioning of streaming videos may also be necessary in order to make them accessible to individuals who are **deaf or hard of hearing**. Additionally, telephones with TTY capability or amplified sound provide accessible communication for these individuals.

Challenges for persons with **cognitive impairments and neurological disabilities** using websites include lack of consistent navigation structure, lack of illustrative non-text materials, and flickering or blinking design elements.

The most common barriers on websites are posed by images or photographs that do not provide identifying text. A screen reader or similar assistive technology cannot read an image. Thus, when images appear on websites without identifying text there is no way for the individual who is blind or has low vision to know what is on the screen. The simple addition of a tag or other description of the image or picture will keep an individual using a screen reader oriented and allow him or her to gain access to the information the image depicts.

Similarly, complex websites often lack navigational headings or links that would make them easy to navigate using a screen reader. With simple instruction, Web designers can easily add those headings. They may also add cues to ensure the proper functioning of keyboard commands. They can also set up their programs to respond to voice interface technology.

III. ACTION STEP 2: UNDERSTAND THE POLICY FRAMEWORK

Our nation's civil rights laws require that covered entities provide **equal opportunity** to qualified individuals with disabilities. Equal opportunity means an opportunity to obtain the same level of performance, or to enjoy the same level of benefits and privileges that are available to similarly situated individuals without disabilities. It is unlawful for the covered entity to use standards, criteria or **methods of administration** that have the purpose or effect of discriminating on the basis of disability. This includes entering into contracts or other arrangements that have a discriminatory effect. In other words, a covered entity is prohibited from doing *indirectly* that which it is prohibited from doing *directly*.

In addition, our nation's civil rights laws include responsibilities for government contractors and federal agencies to take **affirmative action** to employ and advance in employment individuals with disabilities and disabled veterans, including but not limited to recruitment, advertising and job application procedures. These job application procedures include online application systems.

The duty to take affirmative action regarding the employment of qualified persons with disabilities subsumes the duty not to discriminate. Nondiscrimination is the starting point—the first step required of any government contractor or federal agency in fulfilling its affirmative action obligation. However, affirmative action includes much more than nondiscrimination on the basis of disability by an employer; it includes instituting a system of proactive/positive measures/steps that provide qualified persons with disabilities effective opportunity with respect to all employment activities (e.g., recruitment, selection, hiring, placement, promotion, transfer, layoff, termination, compensation and training) at all levels of employment (including the executive level).

The system of proactive/positive measures/steps includes efforts by government contractors or federal agencies to prevent discrimination on the basis of disability before it occurs by periodically carefully and thoroughly evaluating and monitoring their employment practices to identify/detect barriers to employment and, where such barriers are identified, eliminate/remedy them. Affirmative action also includes expanded outreach, recruitment, mentoring, training and management development and creating a work environment that actively welcomes and fosters advancement of qualified persons with disabilities. Affirmative action does not include quotas or granting preferences to individuals with disabilities.[15]

In addition to our nation's civil rights laws, international treaties provide civil rights protections for people with disabilities and international organizations are establishing accessibility standards for websites.

Below is a brief description of key laws, international conventions and international accessibility standards/guidelines that may have applicability to employers:

- The Americans with Disabilities Act
- Sections 501, 503, 504 and 508 of the Rehabilitation Act of 1973, as amended
- Section 255 of the Telecommunications Act of 1996
- Web Content Accessibility Guidelines 2.0
- The UN Convention on the Rights of Persons with Disabilities

A. Federal Civil Rights Laws

The Americans with Disabilities Act (ADA)

Congress adopted the Americans with Disabilities Act (ADA) in 1990.[16] The statute is a comprehensive, broad-reaching mandate to eliminate discrimination on the basis of disability in the areas of employment (Title I), state and local governments (Title II), public accommodations (Title III) and telecommunication relay services for hearing impaired and speech impaired individuals (Title IV). DOJ is responsible for enforcement and implementation of Titles II and III of the ADA, which cover state and local government entities and private businesses, respectively. The Equal Employment Opportunity Commission (EEOC) and the DOJ enforce Title I of the ADA (EEOC for private employers and DOJ for state and local governments). The FCC is responsible for the implementation and enforcement of Title IV.

Under the ADA, it is unlawful for a covered employer to discriminate on the basis of disability against a qualified individual with a disability in regard to any term, condition or privilege of employment, including recruitment, advertising and job application procedures; hiring, promotion and termination; and fringe benefits, apprenticeships, training and sponsored activities, including social and recreational programs. It is also unlawful for a covered employer to participate in a contractual or other arrangement or relationship that has the effect of subjecting the covered employer's own qualified applicant or employee with a disability to discrimination. Unlawful discrimination includes the failure to provide opportunities to participate in or benefit from a privilege of employment that are equal to and as effective and meaningful as those afforded to other individuals and the failure to provide reasonable accommodations unless to do so would result in undue hardship.[17]

Sections 501, 503 and 504 of the Rehabilitation Act of 1973, as amended

Title V of the Rehabilitation Act of 1973, as amended contains a number of provisions designed to safeguard the civil rights of people with disabilities.

- Section 501 of the Rehabilitation Act protects qualified persons with disabilities from employment discrimination by federal departments and agencies. In addition, each federal department or agency is required to develop an affirmative action plan for hiring, placing and advancing qualified individuals with disabilities within the department or agency.[18]

- Under Section 503 of the Rehabilitation Act, covered government contractors are prohibited from engaging in discrimination and are required to take affirmative action to employ and advance in employment qualified persons with disabilities. Certain government contractors are required to develop affirmative action programs.[19]

- Section 504 of the Rehabilitation Act of 1973 prohibits discrimination in federally assisted and federally conducted programs and activities.[20]

Applicability of ADA and Sections 501, 503 and 504 of the Rehabilitation Act to Accessible ICT, including the Internet

When Congress enacted the ADA and Sections 501, 503 and 504 of the Rehabilitation Act, the Internet as we know it today as the venue for information, commerce, services, activities and employment did not exist. For that reason, although the ADA and Sections 501, 503 and 504 guarantee the protection of the rights of individuals with disabilities in a broad array of activities, neither law expressly mentions the Internet or contains requirements regarding developing technologies. Because the Internet was not in general public use when Congress enacted the ADA and the Attorney General and other department or agency officials promulgated regulations to implement it, neither the statute nor the regulations expressly mention it. But the statute and regulations create general rules designed to guarantee people with disabilities equal access to all of the important areas of American civic and economic life. And DOJ made clear, in the preamble to the original 1992 ADA regulations, that the regulations should be interpreted to keep pace with developing technologies. 28 C.F.R. pt. 36, App. B.[21]

DOJ has long taken the position that both state and local government websites *and* the websites of private entities that are public accommodations are covered by the ADA. In other words, the websites of entities covered by both Title II and Title III of the statute are required by law to ensure that their sites are fully accessible to individuals with disabilities.[22]

According to DOJ, there is no doubt that the Internet sites of state and local government entities are covered by Title II of the ADA. Similarly, there is no doubt that the websites of recipients of federal financial assistance are covered by Section 504 of the Rehabilitation Act. DOJ has affirmed the application of these statutes to Internet sites in a technical assistance publication, *Accessibility of State and Local Government Websites to People with Disabilities* (http://www.usdoj.gov/crt/ada/websites2.htm), and in numerous agreements with state and local governments and recipients of federal financial assistance. DOJ's technical assistance publication also provides guidance with simple steps to ensure that government websites have accessible features for individuals with disabilities.

On July 26, 2010, DOJ published an Advance Notice of Proposed Rulemaking regarding *Nondiscrimination on the Basis of Disability; Accessibility of Web Information and Services of State and Local Government Entities and Public Accommodations.*[23] DOJ is soliciting public comments on various issues relating to the establishment of specific regulatory requirements for making services, programs and activities accessible via the Web.

B. Federal Standards and International Guidelines Regarding Accessible ICT

Section 508 of the Rehabilitation Act of 1973, as amended

Section 508 of the Rehabilitation Act requires that when federal departments or agencies develop, procure, maintain or use electronic or information technology, they must ensure that the technology is accessible to and usable by people with disabilities, unless an undue burden would be imposed on the department or agency.[24] Section 508 ensures that federal employees with disabilities have access to, and use of, the information and data they need to do their job, which reduces barriers to job success and upward mobility. The law also helps ensure that members of the public with disabilities have the ability to access government information and services.

Section 508 also requires the Access Board to publish standards setting forth a definition of electronic and information technology and technical and functional performance criteria for such technology. The Access Board is also required to periodically review and, as appropriate, amend the standards to reflect technological advances in electronic and information technology. The Board published the standards on December 21, 2000.[25] On March 22, 2010, the Access Board published in the Federal Register[26] an Advance Notice of Proposed Rulemaking (ANPRM) updating and modernizing the Telecommunications Act Accessibility Guidelines and the Electronic and Information Technology Standards. It issued a second ANPRM on these same issues on December 8, 2011.[27] Through the proposed updates, the Access Board is addressing new technology and seeks to harmonize, to the extent possible, its criteria with other standards and guidelines in order to improve accessibility and facilitate compliance. In particular, the Access Board is seeking to harmonize its Section 508 standards with the World Wide Web Consortium (W3C) Web Content Accessibility Guidelines (WCAG) 2.0 (December 11, 2008). The WCAG 2.0 is described more fully below.

Section 508 uses the procurement process to ensure that ICT acquired by the federal government is accessible to individuals with disabilities. The law also sets up an administrative process under which individuals with disabilities can file a complaint alleging that a federal agency has not complied with the standards.

The standards define the types of technology covered and set forth a minimum level of accessibility. The standards also provide criteria specific to various types of technologies, including:

- Software applications and operating systems
- Web-based information and applications
- Telecommunication products
- Video and multimedia products
- Self-contained, closed products (e.g., information kiosks, calculators, fax machines)
- Desktop and portable computers

The standards provide technical specifications and performance-based requirements, which focus on functional capabilities of covered ICT. According to the Access Board, this approach is designed to recognize the dynamic and continually evolving nature of the ICT as well as the need for clear and specific standards to facilitate compliance. Certain provisions are designed to ensure compatibility with assistive technology people with disabilities use for ICT access, such as screen readers, Braille displays and TTYs.

Section 255 of the Telecommunications Act of 1996

Section 255 of the Telecommunications Act of 1996 requires manufacturers to ensure that telecommunications equipment or customer premises equipment are designed, developed and fabricated to be accessible to and usable by individuals with disabilities when it is readily achievable to do so and readily achievable means easily accomplishable without much difficulty or expense. The Access Board is given responsibility for developing accessibility guidelines for telecommunication equipment and customer premises equipment in conjunction with the FCC.

C. International Accessibility Standards Applicable to Websites[28]

The Web Accessibility Initiative (WAI) of the World Wide Web Consortium (W3C) creates internationally recognized voluntary standards for Web accessibility, including guidelines for Web content, authoring tools, browsers and media players. Specifically, the Web Content Accessibility Guidelines (WCAG) detail how to make Web content and Web applications accessible to individuals with disabilities and older people. The most recent and updated version of the WCAG, the WCAG 2.0, was published in December 2008 and is available at http://www.w3.org/TR/WCAG20/ (last visited June 29, 2010). According to the WAI, the WCAG 2.0 "applies broadly to more advanced technologies; is easier to use and understand; and is more precisely testable with automated testing and human evaluation." *See* WAI, *Web Content Accessibility Guidelines (WCAG) Overview*, available at http://www.w3.org/WAI/intro/wcag.php (last visited June 29, 2010). It is supported by an extensive and expanding library of technical resources for Web developers, including implementation techniques[29] for HTML, CSS, Scripting, Flash and additional Web technologies.

The WCAG 2.0 contains 12 guidelines addressing Web accessibility. Each guideline contains testable criteria for objectively determining if Web content satisfies the guideline. In order for a Web page to conform to WCAG 2.0, the Web page must satisfy applicable "Success Criteria" for these guidelines under one of three conformance levels: A, AA or AAA. The three levels of conformance indicate a measure of accessibility and feasibility. Level A, which is the minimum level of conformance for access, contains criteria that provide basic Web accessibility and that are the most feasible for Web content developers. Level AA, which is the intermediate level for access, contains enhanced criteria that provide more comprehensive Web accessibility and yet are still feasible for Web content developers. Level AAA, which is the maximum level of access, contains criteria that may be less feasible for Web content developers. In fact, WAI does not recommend that Level AAA conformance be required as a general policy for entire websites because it is not possible to satisfy all Level AAA criteria for some content. *See* W3C˙, *Understanding WCAG 2.0: Understanding Conformance* (Dec. 2008), http://www.w3.org/TR/UNDERSTANDING-WCAG20/conformance.html (last visited June 29, 2010).

To give an overview of the types of considerations necessary for making a website accessible, WCAG 2.0 can be briefly paraphrased as follows. Please note that this is not definitive text and should not be used for conformance, only for examples:

WCAG 2.0 At A Glance[30]

Perceivable

   * Provide text alternatives for non-text content.
   * Provide captions and other alternatives for multimedia.
   * Create content that can be presented in different ways,
     including by assistive technologies, without losing meaning.
   * Make it easier for users to see and hear content.

Operable

   * Make all functionality available from a keyboard.
   * Give users enough time to read and use content.
   * Do not use content that causes seizures.
   * Help users navigate and find content.

Understandable

   * Make text readable and understandable.
   * Make content appear and operate in predictable ways.
   * Help users avoid and correct mistakes.

Robust

    * Maximize compatibility with current and future user tools.

D. Applicability of Section 508 and WCAG 2.0 to Private Sector Employers

As explained above, federal civil rights statutes and implementing regulations require that all privileges offered to individuals without disabilities be offered to individuals with disabilities, whether the privileges are provided directly by the company or through contract or other arrangement with another entity. This means individuals with disabilities must be provided an opportunity to take advantage of these privileges that is equal to and as effective as that provided to others.

To date, there are no explicit standards for accessible ICT included in disability rights regulations applicable to private sector employers, state and local governments, and public accommodations. In recognition of the critical importance of accessible ICT in the exercise of these privileges, federal agencies are, however, taking steps to clarify the applicability of these generic policies under the ADA and the Rehabilitation Act to the development, procurement, maintenance and use of accessible ICT in handbooks and other informal guidance and specifying the existence of applicable standards (e.g., Section 508 and WCAG 2.0) that may serve as "safe-harbors."[31] For example, the Department of Health and Human Services (HHS) recently issued *Guidance for Exchange and Medicaid Information Technology (IT) Systems, Version 1.0* (November 2, 1010). With respect to standards for accessibility, this document states[32]:

> "Systems shall include usability features or functions that accommodate the needs of persons with disabilities, including those who use assistive technology. State enrollment and eligibility systems are subject to the program accessibility provisions of Section 504 of the Rehabilitation Act, which include an obligation to provide individuals with disabilities an equal and effective opportunity to benefit from or participate in a program, including those offered through electronic and information technology. At this time, *the Department will consider a recipient's websites, interactive kiosks, and other information systems addressed by Section 508 Standards as being in compliance with Section 504 if such technologies meet those Standards. We encourage states to follow either the 508 guidelines or guidelines that provide greater accessibility to individuals with disabilities. States may wish to consult with latest Section 508 guidelines issued by the US Access Board or W3C's Web Content Accessibility Guidelines (WCAG) 2.0.*" (emphasis added)

With respect to Section 508, it should be noted that all of the Web-based documents developed by the General Services Administration (GSA) for use by federal agencies and vendors are available for use by private sector employers by simply signing in as a guest.[33] With respect to WCAG 2.0, it should be noted that numerous supporting technical and educational documents developed by W3C are also available to the public.[34] Also, as mentioned above, the Access Board recently published an Advance Notice of Proposed Rulemaking that, among other things, is designed to harmonize the Section 508 standards with WCAG 2.0.

E. United Nations Convention on the Rights of Persons with Disabilities

The United Nations General Assembly adopted the Convention on the Rights of Persons with Disabilities in 2006. So far, 148 countries have signed and 100 have ratified this Convention. On July 30, 2009, Susan Rice, President Obama's ambassador to the United Nations, signed the convention at U.N. headquarters in New York. President Obama has pledged to submit the Convention to the Senate for its ratification (advise and consent).

Article 9 of the Convention addresses accessibility. The article specifies that to enable persons with disabilities to live independently and participate fully in all aspects of life, States Parties shall take appropriate measures to ensure to persons with disabilities access, on an equal basis with others, to, among other things, information and communications, including information and communication technologies and systems. More specifically, Article 9 specifies that States Parties shall also take appropriate measures to promote access for persons with disabilities to new information and communications technologies and systems, including the Internet, and promote the design, development, production and distribution of accessible information and communication technologies and systems at an early stage, so that these technologies and systems become accessible at minimum cost.

F. Summary—Importance of Understanding the Policy Framework

As explained above, there are several sets of standards describing how to make ICT, including websites, accessible to individuals with disabilities. Some employers may elect to use the standards that were developed and are maintained by the Access Board in accordance with Section 508 and Section 255 of the Telecommunications Act; whereas other employers may elect to use the WCAG 2.0 guidelines developed and maintained by the WAI of the W3C. For purposes of this Framework, either set of standards/guidelines is appropriate.

The importance for private sector employers of understanding the policy framework is evident by reviewing policy guidance recently issued by the U.S. Department of Education (ED) on the rights of students with disabilities when educational institutions use technology.[35] Using the policy pronouncements in the ED guidance, the following text illustrates the possible application of nondiscrimination standards regarding accessible emerging technologies in the employment context by substituting "company" or "employer" for "schools" and substituting "applicants" and "employees" for "students."

Employers should consider accessibility issues up front when they are deciding whether to create or acquire ICT and when they are planning how the ICT will be used. To that end, employers should include accessibility requirements and analyses as part of their acquisition procedures. Employers should keep in mind their obligation to ensure that applicants and employees with disabilities receive the privileges of employment in an equally effective and equally integrated manner. Among the questions employers should ask are:

- What employment opportunities and privileges does the company provide through the use of the ICT?
- How will the ICT provide these opportunities and privileges?

- Does the ICT exist in a format that is accessible to individuals with disabilities?
- If the ICT is not accessible, can it be modified (see below about additional questions related to modifications), or is there a different technological device available, so that applicants and employees with disabilities can obtain the employment opportunities and privileges in a timely, equally effective and equally integrated manner?

Example: An employer intends to establish a Web mail system so that employees can communicate with each other and with managers; receive important messages from the company (e.g., a message about a health or safety concern); and communicate with individuals outside the company. The company must ensure that the employment benefits, services and opportunities provided to applicants and employees through a Web mail system are provided in an equally effective and equally integrated manner. Before deciding what system to purchase, the company should make an initial inquiry into whether the system is accessible to individuals who are blind or have low vision (e.g., whether the system is compatible with screen readers and whether it gives users the option of using large fonts). If a system is not accessible as designed, the company should take further action to determine whether an accessible product is available, or whether the inaccessible product can be modified so that it is accessible to applicants or employees who are blind or have low vision.

In making the determination about modifications or the purchase and use of different devices available, the questions a company should ask include:

- What employment opportunities and privileges of employment does the company provide through the use of ICT?

- What can the company do to provide applicants and employees with disabilities equal access to the employment privileges or opportunities provided through the use of the ICT?

- How will the employment opportunities and privileges provided to applicants and employees with disabilities compare to the opportunities and privileges that the ICT provides to applicants and employees without disabilities? Three relevant questions are:

  o Are all the employment opportunities and privileges that are available through the use of the ICT equally available to applicants and employees with disabilities through the provision of accommodations or modifications (i.e., do applicants and employees with disabilities have the opportunity to acquire the same information, engage in the same interactions, and enjoy the same services as nondisabled applicants and employees)?

  o Are the employment opportunities and privileges provided to applicants and employees with disabilities in as timely a manner as those provided to applicants and employees without disabilities (i.e., do the time frames under which opportunities and privileges are received by applicants and employees meet the requirement that applicants and employees with disabilities be provided privileges and opportunities in an equally effective and equally integrated manner)?

o Will it be more difficult for applicants and employees with disabilities to obtain the employment opportunities and privileges than it is for applicants and employees without disabilities (i.e., does ease of use for applicants and employees with disabilities meet the requirement that applicants and employees with disabilities be provided privileges and opportunities in an equally effective and equally integrated manner)?

Example: The training department at a company creates an online course that includes instruction, posting of assignments and other course content, and a forum where employees can discuss their course work with the instructor and each other. The instructor would like to incorporate video clips into the course, but is unable to obtain the video clips with audio descriptions. As a modification, the instructor creates separate audio descriptions for each video clip that narrate what is taking place in the video, and places them in a separate section of the online course. The online course includes links that enable persons who use screen readers to bypass the video clips completely and instead listen to the audio descriptions. Here, the use of detailed audio descriptions that are a part of the online course would provide employees with disabilities access to the same opportunities and benefits in an equally effective and equally integrated manner. Companies should also think about whether other accommodations may be needed to provide equal access. For example, an employee who uses a screen reader may need extra time to take an online examination because it may take time for the screen reader to process information displayed on a screen and provide that information to the employee.

IV. ACTION STEP 3: UNDERSTAND THE BUSINESS CASE

In making the business case for developing, procuring, maintaining or using accessible ICT, it is important to recognize that every company is different and every CEO or leading figure within each company is different—every business has a different motivator. Thus, there is a need to provide business leaders and decision makers with different approaches, opportunities and information to determine what constitutes a compelling business case.[36]

Below are examples of factors that can be used to make the business case for developing, procuring, maintaining or using accessible ICT.[37]

- In an era when technology is redefining the workplace and creating a knowledge-based economy that places a high value on communication, collaboration and mobility, accessibility is increasingly essential in doing business.

- Accessible technology addresses the personal needs of all computer users, including those with sensory, physical, and learning and language impairments and age-related limitations, making it easier for companies to empower employees as well as serve customers and engage with partners.

- As technology continues to transform the workplace, demand is growing for accessible technology that can accommodate the needs and preferences of all users.

- Businesses are looking for more effective ways to recruit, empower and retain valuable employees and to increase their efficiency and productivity.

- In today's connected world, technology is at the heart of all of these initiatives. Yet to achieve these goals, companies must make technology and its benefits accessible to the largest possible number of people, regardless of age or ability.

- Accessible technology can make it easier for anyone to see, hear and use a computer. Accessible technology enables people with a wide range of abilities, including those with disabilities, age-related impairments or temporary limitations, in addition to novice computer or Web users, to adjust technology to accommodate their individual visual, dexterity, hearing, learning and language needs and preferences or to modify their personal technology experience.

- Accessibility makes it easier for anyone to see, hear and use a computer and to personalize their computers to meet their own needs and preferences. If technology is truly accessible, it can be adjusted to meet the needs and preferences of people with a wide range of abilities, not just those with disabilities.

- Most people who use computers also benefit from at least some accessibility features and settings that enable them to personalize their computers and work environments.

- Accessible technology helps businesses support workforce diversity, recruit from a larger pool of candidates and enhance team collaboration and communication among all employees, including those with disabilities. For example, accessible technology can facilitate communication between sighted and non-sighted colleagues by providing just-in-time delivery of information without the need for special printing by enabling people who are blind or who have vision impairments to enlarge or customize fonts, choose high-contrast settings for greater visibility, or use a screen reader to access team documents.

- The percentage of the workforce represented by individuals 55 years and older is increasing. As a result of functional limitations that result from aging, these employees will require environmental modifications and/or technology accommodations. As the global population continues to grow older and the number of age-related impairments increases, so does the need for accessible and assistive technology. Many people with age-related disabilities develop those impairments during their working lives.

- Companies that make accessibility a priority are sending a clear message to employees and customers alike that their needs matter. And that can breed satisfaction and loyalty.

- Companies that make accessibility a priority in their business can advertise job openings, requests for proposals, and other opportunities on the websites that are accessible.

- Another powerful consideration for making accessibility a part of any business is the sheer number of people around the world who need accessible technology, could benefit from using it, or choose to use accessible technology for a more comfortable or convenient experience. In the U.S., about one in five residents (54.4 million people) reported having some level of disability in 2005. Microsoft and Forrester Research report that more than half (57%) of computer users in the U.S. could likely benefit from accessible and assistive technology due to mild to moderate difficulties or impairments that interfere with their ability to use a computer or to perform routine tasks.

- In addition, many nations are extending civil rights protections for people with disabilities that encompass accessibility and digital inclusion. Some governments are requiring procurement officials to purchase the most accessible products available, creating economic incentives for businesses to build accessible technology products.

- For companies that manufacture technology, building accessible products and marketing them to people with disabilities can be advantageous. Customers with disabilities and their families, friends and associates represent a trillion dollar market segment. Like other market segments, they purchase products and services from companies that best meet their needs – and when buying technology, that means products that are accessible and usable.

V. ACTION STEP 4: SELF-ASSESS ICT PRACTICES, SET GOALS AND ESTABLISH PRIORITIES FOR IMPROVEMENT

Given the business case and policy mandates for providing accessible ICT, it is advisable for organizations to self-assess their internal and external technologies and set priorities for improvement. This will help facilitate the adoption of formal, written policies, practices and procedures to enhance employment opportunities and privileges of employment for individuals with disabilities through accessible ICT.

Companies should begin by considering all of the ICT used or offered by their organization and then making a list of those platforms, devices and applications. This will position them to evaluate the accessibility of each item on their inventory by considering the user experience of applicants, employees, and customers who have various disabilities; and the ability of their tools and processes to support production of accessible materials. This exercise may involve formal testing of ICT applications with automated accessibility testing tools. (See Section Action Step 8 for helpful testing resources.) Others answers may be found by interviewing existing employees with certain disabilities or conducting informal focus groups. The *Appendix* of this document contains a sample checklist to help guide organizations through the self-assessment exercise.

It is likely that this process will identify ICT in the organization that is not accessible to and usable by people with disabilities. Remedying all of the identified issues at once may not be realistic for some organizations due to cost and legacy IT infrastructure constraints. However, leading companies commit to making all the necessary changes, either immediately or in the longer term. (See Action Step 7 for more information on setting priorities and measurable objectives.)

VI. ACTION STEP 5: ADVANCE CORPORATE POLICIES, PRACTICES AND PROCEDURES

A. Introduction

As mentioned previously, to enhance employment opportunities and privileges of employment for individuals with disabilities through accessible ICT, it is necessary to refine and advance corporate policies, practices and procedures that define the nature and scope of the commitment. Best and promising business practices regarding accessible ICT include the adoption of formal, written policies, practices and procedures. ICT procurements or ICT projects (developed, maintained or used) should comply with specific technical ICT accessible standards and functional performance criteria. These include:

- Software applications and operating systems
- Web-based Intranet and Internet information applications
- Telecommunication products
- Video and multimedia products
- Self-contained, closed products
- Desktop and portable computers

A summary of the current Section 508 regulations applicable to each of these topic areas,[38] which are expected to be replaced with more updated provisions at some future date,[39] follows.

B. Software Applications and Operating Systems

Most of the specifications for software pertain to usability for people with vision impairments. For example, one provision requires alternative keyboard navigation, which is essential for people with vision impairments who cannot rely on pointing devices, such as a mouse. Other provisions address animated displays, color and contrast settings, flash rate and electronic forms, among others.[40] Companies should consider these issues when assessing their internal business applications, such as accounting/financial software, travel booking systems, time entry applications, client databases, customer relationship management (CRM) software and other applications. E-mail systems, instant messaging applications and operating systems also fall into this category. All employees, regardless of disability, should be able to access these systems with or without assistive technologies.

C. Websites

Making websites accessible is neither difficult nor especially costly, and in most cases providing accessibility will not result in changes to the format or appearance of a site.[41] It has been shown that businesses can flourish while producing accessible websites and services. In particular, it has been shown that the WCAG 2.0 guidelines are feasible for simple "Mom and Pop" websites, as well as for complex and dynamically-generated million-page websites.[42]

Web accessibility is about designing and coding pages so as many people as possible can access them effectively and efficiently (universal design). Web accessibility specifically ensures effective and meaningful opportunities for individuals with disabilities to use and interact with the company through the Internet and company website.

According to the Access Board, the criteria for Web-based technology and information included in the Section 508 regulation[43] are based on access guidelines developed by the WAI of the W3C. Many of these provisions ensure access for people with vision impairments who rely on various assistive technology products to access computer-based information, such as screen readers, which translate what's on a computer screen into automated audible output, and refreshable Braille displays. Certain conventions, such as verbal tags or identification of graphics and format devices, like frames, are necessary so that these devices can "read" them for the user in a sensible way. The standards do not prohibit the use of website graphics or animation. Instead, the standards aim to ensure that such information is also available in an accessible format. Generally, this means use of text labels or descriptors for graphics and certain format elements. (HTML code already provides an "Alt Text" tag for graphics that can serve as a verbal descriptor for graphics.) This section also addresses the usability of multimedia presentations, image maps, style sheets, scripting languages, applets and plug-ins, and electronic forms.

Below are examples of areas of focus and practices regarding accessible websites that a company may want to adopt.[44]

Website Home Pages:

- External Internet homepage (mission, activities, information about programs, benefits, information about products and services, publication of resources, employment postings).

- Internal Intranet homepage (mission, activities, programs and benefits, products and services, resources, employment postings).

- Web-based forms (applying for programs and benefits, ordering products or services, feedback, contact information search, publication resource search, filing a complaint, employment search).

- Web-based applications (applying for programs, benefits, ordering products or services, training/learning, travel reservation, time and attendance, recordkeeping/tracking, survey, employment)

Items on a Website:

- Portable document files (.pdfs)
- Multimedia content (video and audio)
- Flash content
- Word processing files
- PowerPoint or other presentation files
- Data tables
- Spreadsheet files
- JavaScript or other scripts

- Java applets
- Blogs (web logs)
- Facebook
- MySpace
- Twitter
- YouTube
- Flickr

Typical barriers (key issues)[45]:

- Images not labeled properly with an alternative text description;
- Inconsistent navigation including poor hypertext link text;
- Inaccessible forms for Web users who are blind and use screen reader software;
- Information validation techniques that cause problems with adaptive technology used by people with disabilities; and
- Information laid out in tables (for example job listings frequently are not coded properly for accessibility).

Create an Accessible Website:

- Determine internal accessibility standards for website. See, for example the standards developed by the Access Board implementing Section 508 of the Rehabilitation Act[46] and the WCAG guidelines.[47] WCAG 2.0 is based on a series of success criteria prioritized into three levels.
- Develop a strategic plan to upgrade the accessibility of current site.
- Use latest Web design software that has inbuilt accessibility prompts.
- Test site's accessibility using available tools (software) and repeat whenever new templates are introduced.[48]
- Consider using experts and individuals with different impairments to audit website.
- Communicate required standards to everyone involved, including vendors/suppliers.
- Ensure that the employees and suppliers who are involved with Web design, maintenance and content development have Web accessibility awareness training.
- Review and update guidelines regularly to incorporate latest developments in Web accessibility, particularly changes to the Section 508 and WCAG guidelines.

D. Telecommunications Products

The provisions in the Section 508 regulations related to telecommunication products are designed primarily to ensure access to people who are deaf or hard of hearing.[49] This includes compatibility with hearing aids, cochlear implants, assistive listening devices and TTYs. TTYs are devices that enable people with hearing or speech impairments to communicate over the telephone; they typically include an acoustic coupler for the telephone handset, a simplified keyboard and a visible message display. One requirement calls for a standard non-acoustic TTY connection point for telecommunication products that allow voice communication but that do provide TTY functionality. Other specifications address adjustable volume controls for output, product interface with hearing technologies, and the usability of keys and controls by people who may have impaired vision or limited dexterity or motor control.

Future updates to the Section 508 regulations are likely to include accessibility considerations for mobile telephones and other wireless devices, which many organizations are issuing to their employees. When choosing these devices, employers should consider the ability to connect an alternative headset to particular devices; whether the keys on the keypad are easily discernible from one another; whether there are adjustable contrast and brightness controls; whether the handset has non-slip grips to prevent the phone from slipping out of the hand; and other display characteristics such as the screen size, adjustability of font sizes and more.[50]

E. Video or Multimedia Products

The Section 508 regulations specify standards for video or multimedia products.[51] Multimedia products involve more than one media and include, but are not limited to, video programs, narrated slide productions and computer generated presentations. Provisions address caption decoder circuitry (for any system with a screen larger than 13 inches) and secondary audio channels for television tuners, including tuner cards for use in computers. The standards also require captioning and audio description for certain training and informational multimedia productions developed or procured by federal agencies. The standards also provide that display or presentation of alternate text or audio descriptions shall be user-selectable, unless permanent.

It is an exemplary practice to apply these same standards to webcasting and video conferencing systems, which many companies are using to help employees communicate from multiple locations. Likewise, these practices should be incorporated into corporate events and activities, where videos can be captioned and audio described, and where presentation materials can be provided in alternative formats.

F. Self-Contained, Closed Products

The Section 508 regulations specify standards regarding self-contained, closed products.[52] This section covers products that generally have embedded software but are often designed in such a way that a user cannot easily attach or install assistive technology. Examples include information kiosks, information transaction machines, copiers, printers, calculators, fax machines and similar types of products. The standards require that access features be built into the system so users do not have to attach an assistive device to it. Other specifications address mechanisms for private listening (handset or a standard headphone jack), touch screens, auditory output and adjustable volume controls, and location of controls in accessible reach ranges.

G. Desktop and Portable Computers

The Section 508 regulations include a section that focuses on keyboards and other mechanically operated controls, touch screens, use of biometric form of identification, and ports and connectors. These standards include ensuring that controls and keys are tactilely discernible without activating the controls or keys; that controls and keys are operable with one hand and do not require tight grasping, pinching or twisting of the wrist; and that the status of all locking or toggle controls or keys is visually discernible and discernible either through touch or sound.[53]

H. Information, Documentation and Support

The Section 508 regulations include standards that address access to all information, documentation and support provided to end users (e.g., federal employees) of covered technologies.[54] This includes user guides, installation guides for end-user installable devices, and customer support and technical support communications. Such information must be available in alternate formats upon request at no additional charge. Alternate formats or methods of communication can include Braille, cassette recordings, large print, electronic text, Internet postings, TTY access, and captioning and audio description for video materials.

I. Online Application Systems

Online application systems have gained popularity because they can be used by companies to streamline the hiring process, including: developing position descriptions, attracting candidates, processing resumes, qualifying job applicants, selecting candidates and making employment offers. Below are descriptions of key components of a typical online application system:[55]

- Website Integration
- Job Posting and Distribution Tools
- Application and Resume Submission
- Communication Between Applicant and Employer
- Resume Extraction and Management
- Candidate Search and Selection Process (including Ranking and Rating)
- Employee Offer Management

Each feature of the online application system should be reviewed to determine whether it is accessible through the company's website.[56] Web accessibility is about designing and coding a company's pages so as many people as possible can access them effectively and efficiently (universal design). Web accessibility specifically ensures effective and meaningful opportunities for individuals with disabilities to use and interact with the company through the Internet and company website and ultimately apply for jobs through the online application system.

- **Website Integration.** Connects the company's website with staffing (i.e., job seekers can submit their resume and complete an online job application). The candidate's information streams directly from the website into Applicant Tracking database.

- **Job Posting and Distribution Tools.** A company is able to access millions of job seekers by automatically posting their job opening from open requisitions to their internal and corporate website and to hundreds of other popular Internet job boards (from big boards to niche and specialized jobboards).

- **Application and Resume Submission**. Candidates can submit their resume electronically and complete an online job application.

- **Communication Between Applicant and Employer.** The company can control when communication with the applicant occurs and what is communicated to the applicant.

- **Resume Extraction and Management.** The company is able to:

  - Add new resumes to a database;
  - Extract detailed information from resume to improve the quality of job order matches;
  - Keep track of actions performed on each applicant until applicant is employed;
  - Import resumes thru e-mails, batch upload or real time; and
  - Configure resumes from different sources as resume repositories.

- **Candidate Search and Selection Process (Including Ranking and Rating)**

  - Companies define their own candidate selection process;
  - Multiple search options provided from the resume repositories to search for best match; and
  - Each candidate's database record and resume can be viewed from the search results and if an e-mail address is provided, candidates can be e-mailed directly from the search results without losing the search results.

- **Employee Offer Management**

  - Once a candidate is initially identified as suitable (ranking and rating), track the candidate through the major stages of the recruiting process, such as:
    - Prescreening
    - Technical skills screening
    - Reference checking
    - Resume submission
    - Interviewing
    - Employment offer sent
    - Offer accepted.
  - Activities for each candidate are recorded so that hiring metrics can be tracked for reporting purposes. Also allows a candidate who has been selected to be converted to an employee.

According to OFCCP, if a company (in the case of Section 503, a government contractor) routinely offers applicants various methods of applying for jobs and all methods of application are treated equally, then an employer may, but need not, ensure that its online application system is fully accessible.[57] But, if a company only uses an online application system to accept applications for employment, it should ensure that potential applicants with disabilities either can use the system or can submit an application in a timely manner through alternative means as an interim measure until the website is made fully accessible. This includes providing a means to contact the company, other than through the online system, to request any reasonable accommodation needed to provide an applicant with a disability an equal opportunity to apply and be considered for the company's jobs. At a minimum, the notice should contain the name of the person to contact and the process for requesting an accommodation.

As explained under Action Step 2, equal opportunity means an opportunity to obtain the same level of performance, or to enjoy the same level of benefits and privileges of employment that are available to similarly situated employees or applicants without disabilities. Factors to consider in determining whether an opportunity is equal to and as effective as that provided to others may include providing the same degree of efficiency, immediacy and convenience.

Below are examples of practices regarding the use of an alternative means of application for employment that a company may want to adopt as interim measures until the website is made fully accessible.[58]

- Determine how the alternative method provides equal opportunity (e.g., provides the same degree of efficiency, immediacy and convenience).
- Advertise the alternative means on the website as an interim measure until the company's website is fully accessible. Emphasize that it is an alternative and is intended only for candidates who cannot use the mainstream system, or who are unreasonably disadvantaged by doing so.
- Highlight the benefits of applying online while assuring applicants that if they cannot apply online they will not face discrimination.
- Provide person-to-person communication. Train and equip HR personnel and "web-help" team to communicate with people with disabilities e.g., textphones, online instant message.
- Keep candidates in the mainstream. Feed candidates back into the mainstream process as quickly as possible.
- Provide for a representative not involved in the employment process to enter information into online application system on their behalf (or outsource this function to their applicant tracking provider).
- **Note:** it is discriminatory to automatically direct <u>all</u> candidates with disabilities to the alternative means.
- If you provide candidates with the option to receive e-mail or text alerts about future vacancies, ensure people with disabilities who have applied <u>offline</u> also have access to this service.

Having an online application system that includes universal design features or that includes many accessibility features for individuals with disabilities does not relieve a contractor of its obligations to provide reasonable accommodations if needed to address the inability of an applicant with a disability to use or access the online system.[59]

Below are examples of practices regarding the provision of reasonable accommodations in the context of online application systems that a contractor may want to adopt.[60]

- Notify the applicant of the right to request a reasonable accommodation if he or she is unable or limited in ability to use or access the online system as a result of disability unless the contractor can demonstrate that the accommodation would impose an undue hardship on the operation of its business.

- Include in a notice an explanation of how to obtain reasonable accommodations via its online application system as well as on its paper applications and job announcements.
- The notices should contain the name of the person to contact and the process for requesting the reasonable accommodation.
- The notices should be prominently displayed and included at the beginning of the online application process.

VII. ACTION STEP 6: ESTABLISH AN INFRASTRUCURE FACILITATING IMPLEMENTATION OF THE POLICIES

To realize and sustain the vision and promises reflected in the formal written policies, practices and procedures (Action Step 5), it is necessary to establish corporate infrastructure and organizational strategies.[61] Best and promising practices regarding the establishment of a corporate infrastructure and organizational strategies implementing an accessible ICT policy include:

- Leadership and Team Approach
- Outsourcing and Procurement
- Training
- Deployment

A. Leadership and Team Approach (Managers Across Divisions)

Leadership at the highest level of business is critical to secure "buy-in" and establish and sustain a corporate-wide culture (not limited to HR) that increases awareness, creates expectations, and enhances commitment to hiring, retention and advancement of persons with disabilities through accessible ICT. The outcome of successful leadership will be a corporate commitment that is pervasive and lasts much longer than the terms of office of one or two corporate leaders that currently support its commitment.[62]

While leadership at the highest levels of a business is necessary to implement and sustain the corporate-wide commitment, it is also critical to recognize that CEOs are policymakers—not policy implementers/enforcers or the people who are actually implementing the work (i.e., people exercising day-to-day responsibility for lines of business where hiring and product development occur). There is a need to secure a network of responsible individuals consisting of policymakers, policy enforcers and policy implementers (different lines of business).[63]

Best practice regarding implementation of an accessible ICT policy includes formalizing a team approach (e.g., an accessibility team comprised of managers across divisions including ADA compliance (ADA Coordinator), HR, ICT (the Chief Information Officer) and the Chief Technology Accessibility Officer, procurement (the Chief Acquisition Officer), education and training, financial and marketing).

B. Outsourcing and Procurement

1. Outsourcing

Many companies choose to outsource all or part of their responsibilities to purchase, develop, procure or maintain ICT including the design and maintenance of websites. As explained under Action Step 2 (Understand the Policy Framework), it is unlawful for companies to participate in a contractual or other arrangement or relationship that has the effect of subjecting the company's own qualified applicants or employees with disabilities to the discrimination prohibited by the regulations. In other words, from a civil rights compliance perspective, the company is responsible for making decisions regarding accessible ICT, not the vendor. Companies must comply with applicable standards. Below are examples of practices regarding outsourcing that a company may want to adopt:[64]

- Conduct a detailed review of all current ICT in line with the company's standards and work with third party suppliers to ensure that the company's ICT is truly accessible.
- Identify who needs to be involved in overcoming barriers, including suppliers and partners (minimizing the risks of outsourcing).
- Insist that job boards on which the company advertises are accessible to people with disabilities.
- Make business reasons for adopting accessible ICT policy when suppliers are invited to tender a proposal.
- Ensure contracts stipulate that suppliers will, when relevant, apply ICT accessibility standards.
- Stipulate who will be responsible for meeting the cost of adjustments.
- Ensure the technology can be used by any employees with disabilities who are involved in recruitment, including those who use assistive technology.
- Give suppliers and business partners a copy of ICT accessibility guidelines.

2. The Procurement Process

A well-written solicitation goes a long way toward ensuring the procurement of accessible ICT.[65] The solicitation should:

- State that the ICT must be accessible.
- Indicate which provisions of the applicable accessible ICT standards apply to purchase.
- Request accessibility information from vendors that respond to solicitation.
- Evaluate received proposals based on responses to accessibility requirements.
- Let vendors know of plans to inspect deliverables based on meeting accessibility requirements (i.e., "acceptance testing").

It is not enough to simply state that the ICT products or services that the company intends to buy must conform to the applicable accessible ICT standards. The procurement officer should identify ICT deliverables covered by the accessible ICT standards and then identify the applicable technical standards; functional performance criteria; and information, documentation and support that apply to each ICT deliverable. It is not permissible to leave it to the vendor to determine if accessible ICT standards are relevant to the solicitation or which standards are applicable. The company is responsible for making these decisions, not the vendor. ICT vendors do not have to comply with anything—companies must comply with applicable standards by procuring products that conform. Vendors who don't make products that conform to the standard are not likely to remain customers of the company for very long until they do.

Creating a compliant ICT solicitation is not difficult if companies include accessibility in procurement planning, identify specifically the provisions that apply to the ICT deliverables being procured, and conduct accessibility market research. These processes cannot be done ad hoc; companies need to have a detailed process for addressing accessible ICT. The BuyAccessible System, a set of free Web-based tools developed by GSA for federal government agencies, provides companies with such a process.[66]

Below is a checklist prepared by GSA for determining compliant solicitations using Section 508 standards. The checklist may be used by private sector employers.

DO THE FOLLOWING:

- Provide a clear statement that an ICT Accessibility Standard (such as Section 508 or WCAG 2.0) does or does not apply to the solicitation.
- Indicate if an exception is being claimed.
- Indicate applicable technical section and provision of the ICT contract deliverables.
- Indicate functional performance criteria applicable to the ICT contract deliverables.
- Indicate which information, documentation and support requirements are applicable to the ICT contract deliverables.
- Specify applicable accessibility factors as evaluation criteria.
- Request accessibility information from the responder.
- Specify applicable accessibility factors as inspection and acceptance criteria.
- Ensure the solicitation and any associated documents and attachments are in an accessible format.

AVOID THE FOLLOWING:

- Do not request that an ICT Accessibility Standard's (such as Section 508 or WCAG 2.0) relevance be determined by the vendor.
- Do not request that the ICT Accessible Standard applicability be determined by the vendor.
- Do not request that ICT Accessibility Standard exceptions be determined by the vendor.
- Do not request an ICT deliverable accessibility certification of compliance (Note: Potential vendors should be asked to provide information on conformance with the stated accessibility factor requirement through a VPAT or GPAT or other documentation. This proof of conformance is not the same as a request for a

certificate of compliance, as there is currently no certifying body for Section 508 accessibility.)

3. *Role of Procurement or Contracting Officials*[67]

As explained above, a compliant solicitation should, among other requirements, articulate the specific accessible ICT standards that apply to the solicitation, not simply include a reference or assurance of compliance with the standards. The Chief Acquisition Officers (CAOs) and Chief Information Officers (CIOs) working with the Chief Accessible Technology Officer, play key collaborative roles in the area of securing the purchase of accessible ICT. CAOs and CIOs, in consultation with the Chief Accessible Technology Officer, should review their procurement and information technology acquisition policies and procedures to validate that the accessible ICT standards are appropriately considered and clearly stated in solicitation documents. CAOs should consider sampling agency procurements that include ICT to ensure they properly specify the appropriate accessible technology standards.

CIOs should have specific obligations including monitoring implementation of the requirement that selected information resource systems or processes facilitate accessibility. Also, CIOs should ensure ICT capital planning and control comply with applicable accessible ICT requirements (i.e., CIOs should instruct agency project managers to review capital planning and investment control documentation for compliance with applicable accessible ICT standards).

1. A procurement or contract official can use the http://www.BuyAccessible.gov website to:

   - Determine if a purchase is subject to the applicable accessible technology standard (such as Section 508 or WCAG 2.0).
   - Find language for the solicitation.
   - Find companies and do market research to buy ICT products or services.
   - Provide documentation for compliance.

2. A procurement or contracting official is responsible for producing specific language for solicitations for ICT products and services. This will include:

   - Basic language regarding Section 508/WCAG 2.0 relevance.
   - Section 508/WCAG 2.0 applicability.
   - Section 508/WCAG 2.0 factors for proposal evaluation.
   - Section 508/WCAG 2.0 criteria for deliverable acceptance.

In addition, the procurement or contracting official will evaluate acquisition deliverables based on generally accepted inspection and/or test methods.

3. The contracting officer—the person with the actual authority to enter into, administer and/or terminate contracts and make related determinations and findings—must understand the requirements of Section 508/WCAG 2.0 relative to procuring ICT

products and services. The contracting officer's responsibilities include:

- Awarding the contract after complying with procurement statutes and regulations (FAR).
- Assisting the requiring official in performing market research for accessible ICT.
- Assisting the appropriate official in drafting specifications and minimum requirements/agency needs when these are need for procurement.
- Ensuring there is competition and that the contracting file shows what happened, as the FAR directs.

C. Accessibility Training for In-House Staff

It is often stated that "people often don't know what they don't know." It is a best and promising practice for companies to extend training and professional development opportunities to all offices, divisions and departments of the company. The company should also expand its base of knowledge and experience through various strategies, demonstrations and newsletters.[68]

The following is a list of staff for which training provided by the Chief Accessible Technology Officer and others would be appropriate:

- ADA coordinators
- Program managers
- Contracting or procurement officers
- Software developers
- Web developers
- Video/multimedia developers
- IT Help desk staff
- Others

The following are possible topics for inclusion in the training:

- ICT accessibility standards
- How to buy accessible ICT
- How to create accessible software
- How to create an accessible website
- How to create accessible electronic forms
- How to create accessible electronic documents
- How to create accessible pdfs
- How to caption video or multimedia
- How to evaluate software, websites and other documents for accessibility

D. Deployment

After a company has purchased or developed accessible ICT, it still has to deploy it throughout the company. To realize the greatest productivity benefits from accessible ICT, companies should:[69]

- Establish a mechanism for centralized expertise and/or payment for assistive technology devices and services.
- Provide for specialized expertise to assess and evaluate an applicant's or employee's need for assistive technology, where necessary and appropriate.
- Train staff in accessibility assessment.
- Enter into strategic partnerships with resources in the community to gain insight, information and expertise regarding accessible ICT, including assistive technology devices and services.
- Provide online resources for learning about specific types of assistive technology products.
- Provide training to employees interested in learning more about how to use assistive technologies and/or accessibility features of various products.
- Involve persons with disabilities in the implementation of accessible ICT policy by establishing a group of employees impacted by the policy.

VIII. ACTION STEP 7: COMMIT TO ACCOUNTABILITY AND CONTINUOUS IMPROVEMENT

The adoption of written policies, practices and procedures and the establishment of corporate infrastructure and organizational systems are necessary components of an ICT accessibility strategic plan. However, ensuring implementation is the end goal. Best business practices include establishing accountability mechanisms and methods for ensuring continuous improvement. Such mechanisms and methods include:

- Appointment of ADA Coordinator, Appointment of Chief Technology Accessibility Officer and Designation of Responsible Managers
- Notice of Policy and Responsible Persons and Offices
- Participation by Individuals with Disabilities and Outside Experts
- Establishment of Measurable Objective and Benchmarks
- Monitoring, Auditing, Reporting and Continuous Improvement
- Prioritizing Areas Needing Improvement

A. Appointment of ADA Coordinator, Chief Technology Accessibility Officer and Designation of Accountable Managers

Designation of authority and assigning responsibility for implementation is a central feature of any implementation effort. Specific examples of strategies and practices regarding the designation of responsibility for implementation that have proven successful include: [70]

(1) Assigning responsibility to specific individual(s) (Chief Technology Accessibility Officer) for implementation of the employer's accessible ICT strategic plan.

(2) Providing these official(s) top management support and, if appropriate, staff to manage the implementation of this strategic plan (affirmative action program), and a reasonable budget.

(3) Identifying the individual on all internal and external communications regarding the company's ICT strategic plan.

(4) Defining the scope of responsibility for implementation to include, among others, the following functions:

- Developing the strategic plan.
- Assisting in the identification of problem areas and in the development of solutions to those problems. For example, making available and keeping updated office addresses, 1-800 telephone numbers and e-mail addresses of the ADA Coordinator(s) and the Chief Accessible Technology Officer. This information should be accessible on the company's "Diversity and Accessibility" page, which should be accessible via a link that is predictable and in an accessible location on the company's website home page, in a format accessible to persons with disabilities.
- Coordinating activities of accessible ICT team members.
- Monitoring the effectiveness of the program on a continuing basis through the development and implementation of an internal audit and reporting system that measures the effectiveness of the program.
- Keeping upper management informed of progress and problems within the company through quarterly reports.
- Providing department managers with a copy of the strategic plan and reviewing the plan with them on an annual basis to ensure knowledge of their responsibilities for implementation of the plan.
- Reviewing the plan with all managers and supervisors at all levels to ensure that the policy is understood and is followed in all activities.
- Auditing the contents of company bulletin boards and Web sites annually to ensure that compliance information is posted and up-to-date.
- Serving as liaison between the company and enforcement agencies.
- Advising managers and supervisors annually of their responsibilities under the company's plan and of their obligations to:
  o Review the company's policy with subordinate managers and supervisors to ensure that they are aware of the policy and understand their obligations; and
  o Assist in the identification of problem areas, formulate solutions, and establish departmental goals and objectives when necessary.

(5) The office(s) or person(s) who have the responsibility for implementing and ensuring compliance with the policy should provide the following services to ensure compliance:

- Assist acquisition officials in preparing accessible ICT language in ICT contracts
- Assist developers in designing software that complies with accessible ICT standards
- Create or repair electronic documents to comply with ICT accessibility standards
- Evaluate websites
- Evaluate software
- Evaluate hardware
- Provide training
- Provide alternate formats

B. Notice of Policy and Responsible Persons and Offices

Post and maintain a copy of the summary description of the ICT accessibility policy on the company's "Diversity and Accessibility" page, which should be accessible via a link that is a predictable and in an accessible location on the company website homepage, in a format accessible to persons with disabilities (i.e., HTML). The summary description is available in alternate formats (such as large print, Braille, audio recording, electronic formats) upon request with reasonable promptness.[71]

C. Participation by Individuals with Disabilities and Outside Experts

In order to enhance accountability and ensure that policies, practices and procedures truly enhance access for applicants and employees with disabilities, it is important to involve people with disabilities and outside efforts in the monitoring and implementation. Best practices include the following:[72]

- Involve people with disabilities at each stage, particularly when testing whether the changes the company has made are successful.

- Develop relationships with experts and learn what other companies are doing.

- Adopt the benchmarks and measures by teams that include experts knowledgeable about accessible and usable electronic and information technology and persons with disabilities.

- Seek input from employees with disabilities using employee surveys and focus groups and discussions with employee diversity and advisory groups regarding the workplace environment. Include persons with disabilities and members of employee resource groups on self-assessment endeavors (e.g., progress and monitoring teams).

D. Measurable Objectives, Benchmarks and Prioritizing Areas Needing Improvement

A "conformance evaluation" or audit of an organization's information and communications technology (ICT) will help companies self-assess the accessibility of their existing technologies and plot a course for improvement. Setting measurable objectives and benchmarks at the onset of this exercise is key to the evaluation and strategic planning process.

While no official method of evaluating ICT accessibility exists, many organizations offer checklists, scorecards and grid-based tracking documents designed to help users evaluate their compliance with Section 508 and other accessibility practices. See the *Appendix* for a partial list of these resources, as well as a sample benchmarking tool.

Some of these technical assessments feature a specific scoring mechanism (e.g., where a red score signifies a lack of Section 508 compliance, yellow indicates partial compliance and green indicates that the technology application is fully accessible).[73] Whether adopting this sort of scoring taxonomy or another, it is a best practice for companies to evaluate each ICT application or platform consistently, and to assign a benchmark "score" to each item at the start of the exercise so that success can be measured.

It is equally important to set measurable objectives by prioritizing ICT applications that need accessibility modifications and setting dates by which improvements will be implemented. Such objectives should be included in the strategic plan and monitored frequently. (See following section.)

As mentioned above, it is highly likely that a company's self-assessment exercise will identify ICT in the organization that is not accessible to and usable by people with disabilities. Remedying all of the identified issues at once may not be realistic for some organizations due to cost and legacy IT infrastructure constraints. However, leading companies commit to making all the necessary changes, either immediately or in the longer term.

Organizing a prioritized list of accessibility errors and problems is an exemplary practice that can make the improvement process more manageable while ensuring the most critical repairs are made first. Priority projects may vary for each company. Some (particularly organizations subject to Sections 503, 504 and 508) may choose to address their regulatory compliance shortcomings first. Others may begin by fixing applications that have been the subject of complaints by users with disabilities. Tasks may also be categorized by ease/level of effort, cost and timing. For example, if a company is about to replace a system, it is usually more logical to ensure that the new application is fully accessible rather than spending time and effort on fixing the legacy system.

The W3C provides guidance on setting priorities for improving website accessibility. It recommends setting a target standard for compliance, for example WCAG 2.0 Level AA (which includes Level A and Level AA). It also explains that in some website retrofitting situations, organizations might define a multi-tiered target with different dates for different levels, for example, "meet WCAG 2.0 Priority Checkpoints in key Web pages within two months, and meet those Checkpoints in all pages within nine months." Additional guidance from W3C includes the following:

Prioritizing by Barriers[74]

- One approach to retrofitting is to fix all of the Level A barriers first, and then fix Level AA barriers later. There are several disadvantages to this approach: having to go back over templates, style sheets and pages multiple times; missing easy-to-do, lower priority items; and getting stuck on a few hard problems, instead of getting lots of easy things done quickly.

- A more effective approach in most cases is to do all of the high impact **and** easy repairs while working on a page, template, style sheet, etc. Then address harder problems later. This approach has several advantages: it usually takes less time overall and many changes are completed quickly. This helps demonstrate a commitment to improving the accessibility of a website as soon as is feasible.

- To plan for retrofitting, consider two parameters for prioritizing the order in which to address accessibility barriers: impact and effort. For each accessibility barrier, determine:

  - **Impact on people with disabilities.** Accessibility barriers are often identified by WCAG 2.0 Checkpoint. Each Checkpoint has a priority that helps determine the impact of a particular accessibility barrier. The actual impact on users of a specific barrier depends on the context of your site. For example, poor color contrast in the main content has high impact on some people with disabilities, whereas in generic footers it may have low impact. Evaluating with users with disabilities also helps determine the impact of accessibility barriers in a website.

  - **Effort required for repair.** The time, cost and skills to repair a barrier varies greatly based on parameters such as the type of repair and the development environment. For example, repairing barriers in footers could require a simple change to the template that is automatically propagated throughout the website, or could require changing every page manually. Repairing missing alternative text equivalents requires knowing the content and understanding how the text is used; whereas, changing a site to effectively use style sheets requires CSS skills.

Prioritizing by Area[75]

- Another aspect of prioritizing retrofitting is determining which areas to work on first. It is usually best to first repair those areas that have the greatest impact on users' experience with a website. Depending on how a site is developed, many barriers may be able to be repaired through:

  - Templates that impact many pages
  - Style sheets that impact many pages
  - Elements that impact many pages, such as navigation bars and scripts
  - Certain pages may have higher priority, such as:
    - The home page, which is often the first entry point to the site
    - The main pages and functionality based on the purpose of the site, including:
      - The path to get there from the common entry points
      - The path to complete transactions, such as ordering products
  - Frequently-used (high traffic) pages and functionality, including the path to get there and complete transactions

E. Monitoring, Auditing, Reporting and Continuous Improvement

Companies understand that "what is measured is what counts."[76] Monitoring, auditing and reporting systems are necessary to ascertain whether current accessible ICT policies, practices and procedures are effective. Specific examples of strategies and practices regarding monitoring, auditing and reporting systems that have proven successful include:[77]

(1) Designing and implementing an audit and reporting system that will:

- Measure the effectiveness of the company's strategic plan.
- Determine the degree to which the company's goals, priorities and objectives have been attained.
- Indicate any need for remedial action.

(2) Where the company's strategic plan is found to be deficient, undertaking necessary action to bring the program into compliance.

(3) Identifying trends and/or issues for making informed decisions on issues needing more attention.

(4) Tracking information that could be used to assess the effectiveness of the accessible ICT policy.

(5) Using the accessible ICT diagnostic evaluation benchmarks and measures to continuously improve and enhance the accessibility and usability of the ICT.

(6) Establish a continuous feedback mechanism that includes a state-of-the-company briefing to the CIO every six months with respect to the accessibility and usability of electronic and information technology, and annually to the CEO with respect to progress made in implementing the affirmative action program, including reaching benchmarks.

(7) Establishing a complaint tracking system. The company might implement a toll-free telephone number (ADA Comment/Complaint Line) and an e-mail address (ADA e-mail address) that are used solely to receive comments and complaints relating to treatment or access to company services, benefits, privileges, advantages, facilities and accommodations for individuals with disabilities. The ADA coordinators should ensure that each call and e-mail with requests or comments requiring a response will receive a response within fourteen days after it is received. The ADA e-mail address can provide an automatically generated acknowledgement of receipt of incoming mail.

(8) Providing for quarterly reports regarding implementation of the company's strategic plan In quarterly reports, identify completion dates and identify the manager who is accountable and responsible for ensuring that the action item is completed in a timely manner. If action items are not completed in a timely manner, company managers should be held accountable for their performance deficiencies.

IX. ACTION STEP 8: IDENTIFY AVAILABLE RESOURCES

There exists a wealth of information and resources to help companies understand accessibility and successfully meet the needs of qualified individuals with disabilities through the development, procurement, maintenance and use of accessible ICT. The final action step a company can take in support of these efforts is to identify and review the educational material available. Below is a sampling of available resources offered by the federal government and key nongovernmental organizations:

A. **General Regulatory Guidance (on Sections 503, 504 and 508 of the Rehabilitation Act of 1973, as amended)**

- Resources for understanding and implementing Section 508 - http://www.section508.gov

- Resources for understanding and complying with Section 503 - http://www.dol.gov/compliance/laws/comp-rehab.htm and http://www.dol.gov/ofccp/regs/compliance/sec503.htm

- Fact sheet for understanding rights under Section 504 from the U.S. Department of Health and Human Resources - http://www.hhs.gov/ocr/civilrights/resources/factsheets/504.pdf

- Frequently asked questions about Section 504 from the U.S. Department of Education - http://www2.ed.gov/about/offices/list/ocr/504faq.html

B. **Understanding the Business Case**

- Information about the business case for hiring people with disabilities Employer Assistance and Resource Network (EARN) - http://www.askearn.org/BusinessCase

- Information about building a business case for hiring people with disabilities from Disability Works - http://www.disabilityworks.org/default.asp?contentID=150

C. **Acquiring and Purchasing Accessible Technology**

- Resources and tools to help in meeting Section 508 requirements, from GSA - http://www.buyaccessible.gov
    - Training for Buyers- https://app.buyaccessible.gov/baw/training/index.html
    - Training for Sellers- http://www.buyaccessible.gov/BAPSD_Training.ppt

- Sample procurement language from the National Center on Disability and Access to Education (NCDAE) - http://ncdae.org/resources/articles/procurement.php#samples

- Guidance on creating Section 508 compliant IT solicitations from the federal government - http://www.section508.gov/docs/GuidanceonCreating508CompliantITSolicitations.pdf

- Equipment demonstrations, equipment loans, a device-exchange program, training and public awareness initiatives from the Maryland Department on Disability MTAP Program - http://www.mdod.maryland.gov/MTAP%20Home.aspx

D. Understanding Assistive Technologies

Vision Impairments

- Overview of assistive technologies from the American Federation for the Blind (AFB) - http://www.afb.org/Section.asp?SectionID=4&TopicID=31

- Article on assistive technologies for individuals with vision impairment from the Arizona Technology Access Program (AzTAP) - http://www2.nau.edu/aztap-p/geriatric/articles/article6.pdf

Dexterity and Mobility Impairments

- Article on assistive technologies for individuals with motor disabilities from WebAIM - http://webaim.org/articles/motor/assistive

- Article on computer use for individuals with motor disabilities from the University of Washington DO-IT Center - http://www.washington.edu/doit/Brochures/Technology/wtmob.html

Hearing Impairments

- Hearing assistive technology overview from the American Speech-Language Hearing Association (ASHA) - http://www.asha.org/public/hearing/treatment/assist_tech.htm
- Hearing assistive technology overview from the Hearing Loss Association of America - http://www.hearingloss.org/content/hearing-assistive-technology

Learning Impairments

- Article on assistive technology for individuals with learning disabilities from the South Carolina Assistive Technology Program - http://www.sc.edu/scatp/ld.htm

- Technology guide for individuals with learning disabilities from LD Online, a leading resource on learning disabilities - http://www.ldonline.org/indepth/technology/techguide.html

Language and Communication Impairments

- Article on assistive technology for individuals with speech impairments from the Arizona Technology Access Program (AzTAP) - http://www2.nau.edu/aztap-p/geriatric/articles/article11.pdf

- Module on expressive language disorders, including relevant assistive technologies, from the Access Project at Colorado State University - http://accessproject.colostate.edu/disability/modules/LD/tut_comm_dis.cfm?display=pg_2

E. Accessibility Testing Tools

- List of web accessibility evaluation tools from the Web Accessibility Initiative (WAI) - http://www.w3.org/WAI/RC/tools/complete

- Accessibility testing and validation information from the Texas Health and Human Services System - http://architecture.hhsc.state.tx.us/myweb/Accessibility/policy_htm/ch6.htm

- Testing website accessibility information from the U.S. Department of Commerce - http://www.osec.doc.gov/webresources/bestpractices/bp3_accessibility_testing.htm

- Flash accessibility course, including testing information, from the U.S Department of Veterans Affairs - http://www.ehealth.va.gov/508/flash

F. Accessibility Training Resources

- Overview of accessibility presentations and training from the Web Accessibility Initiative (WAI) - http://www.w3.org/WAI/training

- Article on web accessibility training from the University of Washington Access IT program - http://www.washington.edu/accessit/articles?73

- Accessible University from the University of Washington - http://www.washington.edu/accesscomputing/AU

- Web accessibility training from Cornell University Information Technology Department - http://www.cit.cornell.edu/policies/accessibility/training.cfm

- Training Services from Knowability, a disability advocacy organization - http://www.knowbility.org/v/training-services

- Training services from the U.S. Access Board - http://www.access-board.gov/training.htm

G. Implementing Accessible Software Applications and Operating Systems

- e-Accessibility Toolkit from ITU and G3iCT - http://www.e-accessibilitytoolkit.org/toolkit/technology_areas/software

41

- Accessible software tutorial from the U.S. Access Board - http://www.access-board.gov/sec508/software-tutorial.htm

- Software accessibility resources from Mozilla - http://www.mozilla.org/access/resources.html

H. Creating Accessible Websites

- Summary of Web Content Accessibility Guidelines - http://www.w3.org/WAI/WCAG20/glance

- Tips for designing accessible websites, including the self-assessment SNAP tool from the Job Accommodation Network (JAN) - http://askjan.org/media/webpages.html

- Accessible design resources from the Employer Assistance and Resource Network (EARN) - http://www.askearn.org/keyword-links.cfm?KeywordID=437

- Summary listing of the Section 508 web accessibility standards - http://section508.gov/index.cfm?fuseAction=stdsdoc#Web

- Article on accessible online chats from WebAIM - http://webaim.org/articles/archives/chats/

I. Creating Accessible Online Job Applications

- Article on making online application processes accessible from the Job Accommodation Network (JAN) - http://askjan.org/corner/vol02iss05.htm

J. Creating Accessible Documents

- Article on creating accessible Microsoft Word documents from WebAIM - http://webaim.org/techniques/word

- Resources for creating accessible documents from the Maine Department of Education's Maine CITE Program - http://www.mainecite.org/awd/accdocs.html

- Guidance on creating accessible PDFs in the federal government - http://www.section508.gov/docs/pdfguidanceforgovernment.pdf

- A "How To" guide on creating accessible documents from the University of California - Hastings http://www.uchastings.edu/accessibility/Creating-Accessible-Documents.pdf

K. Selecting Accessible Telecommunications Products (mobile phones, etc.)

- The Global Accessibility Reporting Initiative, or GARI, a complete database of accessible cell phones - http://www.accesswireless.org/Find/Gari.aspx

- Android review from the American Federation for the Blind (AFB)-
  http://www.afb.org/afbpress/pub.asp?DocID=aw110202

- Resources on accessible telecommunications products from the Trace Center at the University of Wisconsin- http://trace.wisc.edu/world/telecomm/

- Article on accessible telecommunications and office equipment from the Center for Excellence in Disabilities at the University of West Virginia- http://www.cedwvu.org/publications/itbooklet/telecom.php

L. Creating Accessible Videos and Multi-media Products

- Accessible video files tutorial from Texas A&M University- http://webaccess.tamu.edu/education/tutorials/video.html

- Article on creating video and multimedia products for people with sensory impairments from the DO-IT Center at the University of Washington- http://www.washington.edu/doit/Brochures/Technology/vid_sensory.html

- Factsheet on accessible webcasts from the DO-IT Center at the University of Washington- http://www.washington.edu/doit/print.html?ID=1205

- Tutorial on creating accessible videos from Michigan State University- http://webaccess.msu.edu/tutorials/accessible-videos.html

M. Accessibility of Desktop and Portable Computers

- Overview of accessible computer hardware from the Royal National Institute for the Blind Digital Accessibility Team (UK) - http://www.tiresias.org/research/guidelines/hardware.htm

- Article on adapting computers to make them more accessible from Vision Aware- http://www.visionaware.org/making-computers-accessible

N. Accessibility of Social Media

- Article on promising practices on how to address accessibility issues in social media from the Non Profit Technology Network- http://www.nten.org/blog/2011/07/11/addressing-accessibility-social-media

O. Sample Accessibility Evaluation Tools and Checklists

- Article on accessibility evaluation tools from WebAIM - http://webaim.org/articles/tools

- WAVE, a free accessibility evaluation tool from WebAIM - http://wave.webaim.org

- Web accessibility evaluation tools from the Guild of Accessible Web Designers - http://www.gawds.org/show.php?contentid=65

- Accessibility evaluation toolkit and resources from the Victoria government (AU) - http://www.egov.vic.gov.au/victorian-government-resources/manuals-and-toolkits-victoria/accessibility-toolkit/accessibility-toolkit-section-7-a/accessibility-evaluation-tools.html

- SNAP Your Website into Shape tool from the Job Accommodation Network (JAN) - http://askjan.org/bulletins/SNAPTool.htm

- Accessibility scorecard from the Business Taskforce on Accessible Technology - http://www.btat.org/toolkit/amm-scorecard

P. Best Practices

- Accessibility Best Practices Library from the CIO Council of the federal government - http://www.cio.gov/module.cfm/node/bpl

X. APPENDIX: COMPREHENSIVE BENCHMARKING TOOL

As technology continues to transform the workplace, demand is growing for the development, purchase, maintenance, and use of information and communication technology (ICT) that is accessible to and usable by all applicants and employees, including individuals with disabilities. Leading companies recognize that fostering an accessible workplace is the smart thing to do, both from a business standpoint and a legal one.

Action Step 4 and Action Step 7 of the "Framework for Designing and Implementing Accessible Information and Communication Technology (ICT) Strategic Plans" recommend that companies assess their ICT practices, set goals and establish priorities for improvement. Companies are encouraged to consider all of the ICT used or offered by their organization; make a list of those platforms, devices and applications; and evaluate the accessibility of each item on their inventory, along with their general accessible ICT policies and practices. Companies are also advised to review and improve their infrastructure to facilitate the implementation of company policies and their accountability mechanisms and methods for ensuring continuous improvement.

The Benchmarking Tool, which can be expanded into an interactive tool for employers, is a sample instrument that companies might consider using to begin this exercise. It provides a framework for companies to build into their own self-assessment and corporate-wide, comprehensive strategic plan (affirmative action program, where applicable), including accountability mechanisms and methods for ensuring continuous improvement.

More specifically, this Benchmarking Tool is organized in accordance with the key components of an Accessible ICT Strategic Plan:

- Understand the Terminology and Target Population
- Understand the Policy Framework
- Understand the Business Case
- Self-Assess ICT Practices, Set Goals and Establish Priorities for Improvement
- Advance Corporate Policies, Practices and Procedures
- Establish an Infrastructure Facilitating Implementation of the Policies
- Commit to Accountability and Continuous Improvement
- Identify Available Resources

This Benchmakring Tool does not create any new legal requirements or change current legal requirements. Instead, it facilitates the identification of strategies and practices that employers may want to adopt to enhance employment opportunities for qualified individuals with disabilities through accessible ICT.

A. ACTION STEP 1: UNDERSTAND THE TERMINOLOGY AND TARGET POPULATION

The first action step a company may want to take in its efforts to meet the needs of qualified individuals with disabilities through the development, procurement, maintenance and use of accessible ICT is to review the key terms used to describe accessible ICT and the target populations affected by inaccessible ICT systems.

Questions to consider:

1. Does your company identify and define the key terms used to describe and implement the accessible ICT strategic plan?

2. Does your company take steps to identify the target population of applicants and employees affected by inaccessible ICT systems and the target population that could benefit from accessible ICT, including people who:

   - Are blind or have low vision?
   - Have mobility impairments?
   - Are deaf or hard of hearing?
   - Have cognitive impairments and neurological disabilities?
   - Are older workers?

B. ACTION STEP 2: UNDERSTAND THE POLICY FRAMEWORK

Our nation's civil rights laws require that covered entities provide **equal opportunity** to qualified individuals with disabilities. Equal opportunity means an opportunity to obtain the same level of performance, or to enjoy the same level of benefits and privileges that are available to similarly situated individuals without disabilities. It is unlawful for the covered entity to use standards, criteria or methods of administration that have the purpose or effect of discriminating on the basis of disability. This includes entering into contracts or other arrangements that have a discriminatory effect. In other words, a covered entity is prohibited from doing *indirectly* that which it is prohibited from doing *directly*.

In addition, our nation's civil rights laws include responsibilities for government contractors and federal agencies to take **affirmative action** to employ and advance in employment individuals with disabilities and disabled veterans, including but not limited to recruitment, advertising and job application procedures. These job application procedures include online application systems.

Questions to consider:

1. Has your company reviewed the applicable federal and state nondiscrimination and affirmative action laws, regulations and guidelines, including:

   - The Americans with Disabilities Act (ADA)?
   - Sections 503 and 504 of the Rehabilitation Act of 1973, as amended?
   - The United Nations Convention on the Rights of Persons with Disabilities?

2. Has your company reviewed standards that may be used to determine whether ICT is accessible, including:

- Section 508 of the Rehabilitation Act of 1973, as amended?
- Section 255 of the Telecommunications Act of 1996?
- Web Content Accessibility Guidelines (WCAG) 2.0 developed by the Web Accessibility Initiative (WAI) of the World Wide Web Consortium (W3C)?
- The United Nations Convention on the Rights of Persons with Disabilities?

C. ACTION STEP 3: UNDERSTAND THE BUSINESS CASE

In making the business case for developing, procuring, maintaining or using accessible ICT, it is important to recognize that every company is different and every CEO within each company is different—every business has a different motivator. Thus, there is a need to provide business leaders and decision makers with different approaches, opportunities and information to determine what constitutes a compelling business case.

Questions to consider:

1. Does your company understand the business case for developing, procuring, maintaining and using accessible ICT? In other words, does your company identify and understand the benefits of accessible ICT, such as:
   - Enhanced communication, collaboration and efficiency?
   - Reductions in costs associated with the provisions of more cost-effective accommodations to address the needs and preferences of users?
   - Improved recruitment and retention?
   - Achievement of diversity?
   - The image of corporate responsibility?

D. ACTION STEP 4: SELF-ASSESS ICT PRACTICES, SET GOALS AND ESTABLISH PRIORITIES FOR IMPROVEMENT

Given the business case and policy mandates for providing accessible information and communications technology (ICT), it is advisable for organizations to self-assess their internal and external technologies. This will help facilitate the adoption of formal, written policies, practices and procedures to enhance employment opportunities and privileges of employment for individuals with disabilities through accessible ICT.

Questions to consider:

1. Has your company undertaken a self-assessment of all of the ICT used or offered by the company and made a list of those platforms, devices and applications, and the ability of their tools and processes to support production of accessible materials?

2. Has your company evaluated the accessibility of each item on the inventory by considering the user experience of applicants, employees and customers who have various disabilities, including formal testing of ICT applications with automated accessibility testing tools?

3. Has your company interviewed existing employees with certain disabilities or conducted informal focus groups?

E. ACTION STEP 5: ADVANCE CORPORATE POLICIES, PRACTICES AND PROCEDURES

Adopting Corporate Policies

To enhance employment opportunities and privileges of employment for individuals with disabilities through accessible ICT, it is necessary to refine and advance corporate policies, practices and procedures that define the nature and scope of the commitment. Best and promising business practices regarding accessible ICT include the adoption of formal, written policies, practices and procedures.

Questions to Consider:

1. Has your company established a formal, written, company-wide policy and standards regarding accessible ICT?

2. Does your organization's formal, written policy identify the office and/or person responsible for implementing and complying with accessible ICT policy and standards?

3. Does your company's formal, written policy define roles and responsibilities for ensuring ICT accessibility, including the Chief Technology Accessibility Officer, procurement officers, contracting officers, application developers and Web content managers/authors?

Applying Accessibility Standards

There are several sets of standards describing how to make ICT, including websites, accessible to individuals with disabilities. Some employers may elect to use the standards that were developed and are maintained by the Access Board in accordance with Section 508 of the Rehabilitation Act, as amended and Section 255 of the Telecommunications Act; whereas other employers may elect to use the WCAG 2.0 guidelines developed and maintained by the Web Accessibility Initiative (WAI) of the World Wide Web Consortium (W3C). For purposes of this Benchmarking Tool, either set of standards/guidelines is appropriate. ICT procurements or ICT projects (developed, maintained or used) should comply with specific technical ICT accessible standards and functional performance criteria:

- o Software applications and operating systems
- o Web-based Intranet and Internet information applications
- o Telecommunication products
- o Video and multimedia products
- o Self-contained, closed products
- o Desktop and portable computers

- **Software Applications and Operating Systems**

Most of the specifications for software pertain to usability for people with vision impairments. For example, one provision requires alternative keyboard navigation, which is essential for people with vision impairments who cannot rely on pointing devices, such as a mouse. Other provisions address animated displays, color and contrast settings, flash rate and electronic forms, among others.

Questions to consider:

1. Has your company considered ICT accessibility when assessing its internal business applications, such as:
   - Accounting/financial software?
   - Travel booking systems?
   - Time entry applications?
   - Client databases?
   - Customer relationship management (CRM) software and other software applications?
   - E-mail systems?
   - Content management systems (CMS) and other web authoring tools?
   - Instant messaging applications?
   - Operating systems?

2. Are your company's internal and external e-mails accessible? That is, if your e-mails are formatted in HTML, do you follow website guidelines and offer text alternatives and tagged images?

3. If your company uses social media applications, are they accessible? Do you caption any videos that you post? Do you add descriptive titles to photographs and images that you post? Are you descriptive in your status updates? And do you offer your social media content in an alternate location? (For instance, by posting it on your website, blog, through an accessible video player and in your company's online photo gallery.)

4. Are your company word processing software products accessible? Have you trained your employees on ways to ensure that every document they produce can be read by everyone?

5. Can all employees, regardless of disability, access these systems with or without assistive technologies?

- **Websites**

Web accessibility is about designing and coding pages so as many people as possible can access them effectively and efficiently (universal design). Web accessibility specifically ensures effective and meaningful opportunities for individuals with disabilities to use and interact with the company through the Internet and company website.

According to the Access Board, the criteria for Web-based technology and information included in the Section 508 regulation are based on access guidelines developed by the Web Accessibility Initiative (WAI) of the World Wide Web Consortium (W3C).

Questions to consider:

1. Does the company comply with the applicable standards/guidelines and address typical barriers, including:

   - Images not labeled properly with an alternative text description?
   - Inconsistent navigation including poor hypertext link text?
   - Inaccessible forms for Web users who are blind and use screen reader software?
   - Information validation techniques that cause problems with adaptive technology used by disabled people?
   - Information laid out in tables? (For example job listings frequently are not coded properly for accessibility.)

2. When applying standards/guidelines that address typical barriers in creating the company's website(s), are the following components considered:

   For Web Home Pages:

   - External Internet home page (mission, activities, information about programs, benefits, information about products and services, publication of resources, employment postings)?

   - Internal Intranet home page (mission, activities, programs and benefits, products and services, resources, employment postings)?

   - Web-based forms (applying for programs and benefits, ordering products or services, feedback, contact information search, publication resource search, filing a complaint, employment search)?

   - Web-based applications (applying for programs, benefits, ordering products or services, training/learning, travel reservation, time and attendance, recordkeeping/tracking, survey, employment)?

For Items on a Website:

   - Portable document files (.pdfs)?
   - Multimedia content (video and audio)?
   - Flash content?
   - Word processing files?

- Microsoft PowerPoint files?
- Data tables?
- Spreadsheet files?
- JavaScript or other scripts?
- Java applets?
- Blogs (web logs)?
- Facebook?
- MySpace?
- Twitter?
- YouTube?
- Flickr?

3. When creating an accessible website, does the company take into consideration the following best practices:

- Determine internal accessibility standards for website? (For example, the standards developed by the Access Board implementing Section 508 of the Rehabilitation Act, as amended and the World Wide Web Consortium (W3C) Web Content Accessibility Guidelines (WCAG). WCAG 2.0 is based on a series of success criteria prioritized into three levels.)
- Develop a strategic plan to upgrade the accessibility of current site?
- Use latest Web design software that has inbuilt accessibility prompts?
- Test site's accessibility using available tools (software) and repeat whenever new templates are introduced?
- Consider using experts and individuals with different impairments to audit website?
- Communicate required standards to everyone involved, including vendors/suppliers?
- Ensure that the employees and suppliers who are involved with Web design, maintenance and content development have Web accessibility awareness training?
- Review and update guidelines regularly to incorporate latest developments in Web accessibility, particularly changes to the Section 508 and WCAG guidelines?

- **Telecommunications Products**

The Section 508 regulations include provisions related to telecommunication products, including telephones, mobile devices and other wireless devices.

Questions to Consider:

1. Does the company comply with the accessibility standards applicable to telecommunications products, including accessibility standards for people who are deaf or hard of hearing, such as compatibility with hearing aids, cochlear implants, assistive listening devices and TTYs and the requirement calling for a standard non-acoustic TTY connection point for telecommunication products that allow voice communication but

that do provide TTY functionality, and other specifications address adjustable volume controls for output, product interface with hearing technologies, and the usability of keys and controls by people who may have impaired vision or limited dexterity or motor control?

2. Future updates to the Section 508 regulations are likely to include accessibility considerations for mobile telephones and other wireless devices, which many organizations are issuing to their employees. When choosing these devices, is the company considering the following issues:

   - Ability to connect an alternative headset to particular devices?
   - Whether the keys on the keypad are easily discernible from one another?
   - Whether there are adjustable contrast and brightness controls?
   - Whether the handset has non-slip grips to prevent the phone from slipping out of the hand?
   - Other display characteristics such as the screen size, adjustability of font sizes and more?

- **Video or Multimedia Products**

The Section 508 regulations specify standards for video or multimedia products. Multimedia products involve more than one media and include, but are not limited to, video programs, narrated slide productions and computer generated presentations.

Questions to Consider:

1. Does the company comply with Section 508 or other recognized standards applicable to video or multimedia products such as:

   - Caption decoder circuitry (for any system with a screen larger than 13 inches) and secondary audio channels for television tuners, including tuner cards for use in computers?
   - Captioning and audio description for certain training and informational multimedia productions developed or procured by federal agencies?
   - Display or presentation of alternate text or audio descriptions that are user-selectable unless permanent?

2. Does the company apply these same standards to webcasting and video conferencing systems, which many companies are using to help employees communicate from multiple locations?

3. Does the company incorporate these standards for corporate events and activities, where videos can be captioned and audio described, and where presentation materials can be provided in alternative formats?

- **Self-Contained, Closed Products**

  The Section 508 regulations specify standards regarding self-contained, closed products. This section covers products that generally have embedded software but are often designed in such a way that a user cannot easily attach or install assistive technology. Examples include:

  - Information kiosks;
  - Information transaction machines;
  - Copiers;
  - Printers;
  - Calculators;
  - Fax machines; and
  - Similar types of products.

  Questions to Consider:

  1. Does the company ensure that these products used by company applicants and employees comply with the standards requiring that access features be built into the system so users do not have to attach an assistive device to it?

  2. Do these products comply with other specifications that address mechanisms for private listening (handset or a standard headphone jack), touch screens, auditory output and adjustable volume controls, and location of controls in accessible ranges?

- **Desktop and Portable Computers**

  The Section 508 regulations include a section that focuses on keyboards and other mechanically operated controls, touch screens, use of biometric form of identification, and ports and connectors. These standards include ensuring that controls and keys are tactilely discernible without activating the controls or keys; that controls and keys are operable with one hand and do not require tight grasping, pinching or twisting of the wrist; and that the status of all locking or toggle controls or keys is visually discernible, and discernible either through touch or sound.

  Questions to Consider:

  1. Do the desktop and portable computers, including laptops and tablets, purchased or leased by the company comply with these standards?

Information, Documentation and Support

The Section 508 regulations include standards that address access to all information, documentation and support provided to end users (e.g., federal employees) of covered technologies. This includes:

- User guides;
- Installation guides for end-user installable devices; and
- Customer support and technical support communications.

Questions to Consider:

1. Does the company comply with these standards, including requirements that such information must be available in alternate formats upon request at no additional charge, including:

   - Braille?
   - Cassette recordings?
   - Large print?
   - Electronic text?
   - Internet postings?
   - TTY access?
   - Captioning and audio description for video materials?

- **Online Application Systems**

  Online application systems have gained popularity because they can be used by companies to streamline the hiring process, including: developing position descriptions, attracting candidates, processing resumes, qualifying job applicants, selecting candidates and making employment offers.

  Questions to Consider:

  1. Are the following features of the online application system determined to be accessible through the company's website:

     - Website Integration?
     - Job Posting and Distribution Tools?
     - Application and Resume Submission?
     - Communication between Applicant and Employer?
     - Resume Extraction and Management?
     - Candidate Search and Selection Process (including Ranking and Rating)?
     - Employee Offer Management?

  2. Can potential applicants with disabilities either use your online job application system or submit an application in a timely manager through alternative means as an interim measure until the website is made fully accessible?

  3. If the company uses an alternative method of application as an interim measure until the onsite application is fully accessible, does the alternative method satisfy the following criteria?

- Provides equal opportunity (e.g., provides the same degree of efficiency, immediacy and convenience).
- Is advertised as an interim measure until the company's website is fully accessible, with emphasis that it is an alternative and is intended only for candidates who cannot use the mainstream system, or who are unreasonably disadvantaged by doing so.
- Highlights the benefits of applying online while assuring applicants that if they cannot apply online they will not face discrimination.
- Provides person-to-person communication. Train and equip HR personnel and "web-help" team to communicate with people with disabilities (e.g., textphones, online instant message).
- Keeps candidates in the mainstream or feeds candidates back into the mainstream process as quickly as possible.
- Provides for a representative not involved in the employment process to enter information into online application system on their behalf (or outsource this function to their applicant tracking provider).
- **Note:** it is discriminatory to automatically direct <u>all</u> candidates with disabilities to the alternative means. If you provide candidates with the option to receive e-mail or text alerts about future vacancies, ensure people with disabilities who have applied <u>offline</u> also have access to this service.

4. Does the company use the following practices regarding the provision of reasonable accommodations in the context of online application systems?

- Notify the applicant of the right to request a reasonable accommodation if he or she is unable or limited in ability to use or access the online system as a result of disability unless the employer can demonstrate that the accommodation would impose an undue hardship on the operation of its business.
- Include in a notice an explanation of how to obtain reasonable accommodations via its online application system as well as on its paper applications and job announcements.
- The notices should contain the name of the person to contact and the process for requesting the reasonable accommodation.
- The notices should be prominently displayed and included at the beginning of the online application process.

- **Note:** Having an online application system that includes universal design features or that includes many accessibility features for individuals with disabilities does not relieve a federal contractor of its obligations to provide reasonable accommodations if needed to address the inability of an applicant with a disability to use or access the online system.

4. Does your online application site display the company's equal employment opportunity (EEO) policy statement?

F. ACTION STEP 6: ESTABLISH AN INFRASTRUCURE FACILITATING IMPLEMENTATION OF THE POLICIES

To realize and sustain the vision and promises reflected in the formal written policies, practices and procedures (Action Step 5), it is necessary to establish corporate infrastructure and organizational strategies. Best and promising practices regarding the establishment of a corporate infrastructure and organizational strategies implementing an accessible ICT policy include:

- Leadership and Team Approach
- Outsourcing and Procurement
- Training
- Deployment

Leadership and Team Approach (Managers Across Divisions)

Questions to Consider:

1. Does the company's leadership take action steps to secure "buy-in" and establish and sustain a corporate-wide culture (not limited to HR) that increases awareness, creates expectations, and enhances commitment to hiring, retention and advancement of person with disabilities through accessible ICT?

2. Does the company have a written policy signed and circulated to managers and posted on the website enunciating a policy regarding accessible ICT?

3. Is the company taking action steps to secure a network of responsible individuals consisting of corporate policymakers, policy enforcers and people doing the work (different lines of business) to implement the accessible ICT policy?

4. Is the company implementing a formalized a team approach (e.g., an accessibility team comprised of managers across divisions including ADA compliance (ADA Coordinator), HR, ICT (the Chief Information Officer (CIO) and the Chief Technology Accessibility Officer, procurement (the Chief Acquisition Officer), education and training, financial and marketing?

5. Is the CEO briefed on a regular basis regarding the continuing progress in implementing the accessible ICT strategic plan based on written reports prepared by responsible managers?

Outsourcing and Procurement

Questions to Consider:

Outsourcing

1. Does the company:

   - Conduct a detailed review of all current ICT in line with the company's standards and work with third party suppliers to ensure that the company's ICT is truly accessible?
   - Identify who needs to be involved in overcoming barriers, including suppliers and partners (minimizing the risks of outsourcing)?
   - Insist that job boards on which the company advertises are accessible to people with disabilities?
   - Make business reasons for adopting accessible ICT policy when suppliers are invited to tender a proposal?
   - Ensure contracts stipulate that suppliers will, where relevant, apply ICT accessibility standards?
   - Stipulate who will be responsible for meeting the cost of adjustments?
   - Ensure the technology can be used by any employees with disabilities who are involved with managing recruitment, including those who use assistive technology?
   - Give suppliers and business partners a copy of ICT accessibility guidelines?

The Procurement Process

1. When procuring IT products for your company, do you have a written policy and/or detailed RFP language mandating that those products be accessible? Does the solicitation:

   - State that the ICT must be accessible?
   - Indicate which provisions of the applicable accessible ICT standards apply to purchase?
   - Request accessibility information from vendors that respond to solicitation?
   - Evaluate received proposals based on responses to accessibility requirements?
   - Let vendors know of plans to inspect deliverables based on meeting accessibility requirements (i.e., "acceptance testing")?

2. Does your company evaluate received proposals based on their responses to accessibility requirements?

3. Before signing off on newly procured ICT for your organization, does the company conduct "acceptance testing" on products and services received from outside vendors to ensure their accessibility?

The following checklist was prepared by General Services Administration (GSA) for determining compliant solicitations using Section 508 standards. The checklist may be used by private sector employers.

DO THE FOLLOWING:

- Provide a clear statement that an ICT Accessibility Standard (such as Section 508 or WCAG 2.0) does or does not apply to the solicitation
- Indicate if an exception is being claimed
- Indicate applicable technical section and provision of the ICT contract deliverables
- Indicate functional performance criteria applicable to the ICT contract deliverables
- Indicate which information, documentation and support requirements are applicable to the ICT contract deliverables
- Specify applicable accessibility factors as evaluation criteria
- Request accessibility information from the responder
- Specify applicable accessibility factors as inspection and acceptance criteria
- Ensure the solicitation and any associated documents and attachments are in an accessible format

AVOID THE FOLLOWING:

- Do not request that an ICT Accessibility Standard's (such as Section 508 or WCAG 2.0) relevance be determined by the vendor
- Do not request that the ICT Accessible Standard applicability be determined by the vendor
- Do not request that ICT Accessibility Standard exceptions be determined by the vendor
- Do not request an ICT deliverable accessibility certification of compliance

  Note: Potential vendors should be asked to provide proof of conformance with the stated accessibility factor requirement through a VPAT or GPAT or other documentation. This proof of conformance is not the same as a request for a certificate of compliance, as there is currently no certifying body for Section 508 accessibility.

1. **Role of Procurement Officials**

Does the procurement officer:

- Identify ICT deliverables covered by the accessible ICT standards and then identify the applicable technical standards; functional performance criteria; and information, documentation and support that apply to each ICT deliverable?
- Determine if a purchase is subject to the applicable accessible technology standard (such as Section 508 or WCAG 2.0)?
- Find language for the solicitation?
- Find companies and do market research to buy ICT products or services?
- Provide documentation for compliance?

- Produce specific language for solicitations for ICT products and services, including:

  - Basic language regarding Section 508/WCAG 2.0 relevance?
  - Section 508/WCAG 2.0 applicability?
  - Section 508/WCAG 2.0 factors for proposal evaluation?
  - Section 508/WCAG 2.0 criteria for deliverable acceptance?

- Evaluate acquisition deliverables based on generally accepted inspection and/or test methods?

2. Accessibility Training for In-House Staff

- Does the company extend training and professional development opportunities to all offices, divisions and departments of the company?

- Does the company expand its base of knowledge and experience through various strategies, demonstrations and newsletters?

- Does the company's Chief Accessible Technology Officer provide ICT accessibility training to staff, including, where appropriate, such as:

  - ADA coordinators?
  - Program managers?
  - Contracting officers?
  - Software developers?
  - Web developers?
  - Video/multimedia developers?
  - IT Help desk staff?
  - Others?

- Does the training include the following topics:

  - ICT accessibility standards?
  - How to buy accessible ICT?
  - How to create accessible software?
  - How to create an accessible website?
  - How to create accessible electronic forms?
  - How to create accessible electronic documents?
  - How to create accessible pdfs?
  - How to caption video or multimedia?
  - How to evaluate software, websites and other documents for accessibility?

3. Deployment

- Does your company have a centralized method of providing assistive technology to employees (e.g., a centralized accommodation fund)? Has the company maximized the implementation of ICT accessibility policies and implemented the following best practices:

  - Establish a mechanism for centralized expertise and/or payment for assistive technology devices and services?
  - Provide for specialized expertise to assess and evaluate an applicant's or employee's need for assistive technology, where necessary and appropriate?
  - Train staff in accessibility assessment?
  - Enter into strategic partnerships with resources in the community to gain insight, information and expertise regarding accessible ICT, including assistive technology devices and services?
  - Provide online resources for learning about specific types of assistive technology products?
  - Train all employees (individuals with and without disabilities) so that everyone understands and can use the accessibility features in any technology purchased or developed in-house?
  - Involve persons with disabilities in the implementation of accessible ICT policy by establishing a group of employees impacted by the policy?

G. ACTION STEP 7: COMMIT TO ACCOUNTABILITY AND CONTINUOUS IMPROVEMENT

The adoption of written policies, practices and procedures and the establishment of corporate infrastructure and organizational systems are necessary components of an ICT accessibility strategic plan. However, ensuring implementation is the end goal. Best business practices include establishing accountability mechanisms and methods for ensuring continuous improvement.

Appointment of ADA Coordinator, Chief Technology Accessibility Officer and Designation of Accountable Managers

Questions to Consider:

1. Does the company assign responsibility to specific individual(s) (Chief Technology Accessibility Officer) for implementation of the employer's accessible ICT strategic plan?

2. Does the company provide these official(s) top management support and, if appropriate, staff to manage the implementation of this strategic plan (affirmative action program), and a reasonable budget?

3. Does the company identify the responsible individual on all internal and external communications regarding the company's ICT strategic plan?

4. Does the company define the scope of responsibility for implementation to include, among others, the following functions:

   a) Developing the strategic plan?
   b) Assisting in the identification of problem areas and in the development of solutions to those problems. For example, making available and keeping updated office addresses, 1-800 telephone numbers and e-mail addresses of the ADA Coordinator(s) and the Chief Accessible Technology Officer. This information should be accessible on the company's "Diversity and Accessibility" page which is accessible via a link that is predictable and accessible location on the company's website home page, in a format accessible to persons with disabilities?
   c) Coordinating activities of accessible ICT team members?
   d) Monitoring the effectiveness of the program on a continuing basis through the development and implementation of an internal audit and reporting system that measures the effectiveness of the program?
   e) Keeping upper management informed of progress and problems within the company through quarterly reports?
   f) Providing department managers with a copy of the strategic plan and reviewing the plan with them on an annual basis to ensure knowledge of their responsibilities for implementation of the plan?
   g) Reviewing the plan with all managers and supervisors at all levels to ensure that the policy is understood and is followed in all activities?
   h) Auditing the contents of company bulletin boards and Web sites annually to ensure that compliance information is posted and is up-to-date?
   i) Serving as liaison between the company and enforcement agencies; and
   j) Advising managers and supervisors annually of their responsibilities under the company's plan and of their obligations to:
      (i) Review the company's policy with subordinate managers and supervisors to ensure that they are aware of the policy and understand their obligations?
      (ii) Assist in the identification of problem areas, formulate solutions, and establish departmental goals and objectives when necessary?

5. Does the company specify the services to be provided by the office(s) or person(s) responsible for implementing and ensuring compliance with the policy, including:

   a) Assist acquisition officials in preparing accessible ICT language in ICT contracts?
   b) Assist developers in designing software that complies with accessible ICT Standards?
   c) Create or repair electronic documents to comply with ICT accessible standards?
   d) Evaluate websites?
   e) Evaluate software?
   f) Evaluate hardware?
   g) Provide training?
   h) Provide alternate formats?

Notice of Policy and Responsible Persons and Offices

Questions to Consider:

1. Does the company post and maintain a copy of the summary description of the ICT accessibility policy on the company's "Diversity and Accessibility" page?

2. Is the "Diversity and Accessibility" page accessible via a link that is a predictable and in an accessible location on the company website home page, in a format accessible to persons with disabilities (i.e., HTML)?

3. Is the summary description available in alternate formats (such as large print, Braille, audio recording, electronic formats) upon request with reasonable promptness?

Participation by Individuals with Disabilities and Outside Experts

Questions to Consider:

1. Does the company involve people with disabilities at each stage of the strategic plan, particularly when testing whether the changes the company has made are successful?

2. Does the company develop relationships with experts and learn what other companies are doing?

3. Does the company adopt the benchmarks and measures by teams that include experts knowledgeable about accessible and usable electronic and information technology and persons with disabilities?

4. Does the company seek input from employees with disabilities using employee surveys and focus groups and discussions with employee diversity and advisory groups regarding the workplace environment?

5. Does the company include persons with disabilities and members of employee resource groups on self-assessment endeavors (e.g., progress and monitoring teams)?

Measurable Objectives, Benchmarks and Prioritizing Areas Needing Improvement

Questions to Consider:

1. Does the company conduct a "conformance evaluation" or audit of an organization's information and communications technology (ICT) to help the company self-assess the accessibility of their existing technologies and plot a course for improvement?

2. Does the company use a specific scoring mechanism (e.g., where a red score signifies a lack of Section 508 compliance, yellow indicates partial compliance and green indicates that the technology application is fully accessible)?

3. Does the company assign a benchmark "score" for each ICT application or platform at the start of the exercise so that success can be measured?

4. Does the company set measurable objectives by prioritizing ICT applications that need accessibility modifications and setting dates by which improvements will be implemented? Are these objectives included in the strategic plan and monitored frequently?

5. Does the company organize a prioritized list of accessibility errors and problems that can make the improvement process more manageable while ensuring the most critical repairs are made first?

6. Does the company use the W3C guidance on setting priorities for improving website accessibility or other guidance?

Monitoring, Auditing, Reporting and Continuous Improvement

Questions to Consider:

1. Has the company designed and implement an audit and reporting system that will:

    a. Measure the effectiveness of the company's strategic plan?
    b. Determine the degree to which the company's goals, priorities and objectives have been attained?
    c. Indicate any need for remedial action?

2. Where the company's strategic plan is found to be deficient, does the company undertake necessary action to bring the program into compliance?

3. Does the company identify trends and/or issues for making informed decisions on issues needing more attention?

4. Does the company track information that could be used to assess the effectiveness of the accessible ICT policy?

5. Does the company use the accessible ICT diagnostic evaluation benchmarks and measures to continuously improve and enhance the accessibility and usability of the ICT?

6. Does the company establish a continuous feedback mechanism that includes a state-of-the-company briefing to the CIO every six months with respect to the accessibility and usability of electronic and information technology, and annually to the CEO with respect to progress made in implementing the affirmative action program, including reaching benchmarks?

7. Does the company establish a complaint tracking system, for example implement a toll free telephone number (ADA Comment/Complaint Line) and an e-mail address (ADA e-

mail address) that are used solely to receive comments and complaints relating to treatment or access to company services, benefits, privileges, advantages, facilities and accommodations for individuals with disabilities?

8. Does the company provide for quarterly reports regarding implementation of the company's strategic plan? *Note: In quarterly reports, identify completion dates and identify the manager who is accountable and responsible for ensuring that the action item is completed in a timely manner. If action items are not completed in a timely manner, company managers should be held accountable for their performance deficiencies.*

H. ACTION STEP 8: IDENTIFY AVAILABLE RESOURCES

Questions to Consider:

1. Has the company identified resources that can be used by company officials, managers, and other employees to carry out the roles and responsibilities specified in the strategic plan, including:

   - Regulations, guidelines and policy briefs related to the ADA and Section 503 and 504 of the Rehabilitation Act?
   - Section 508 of the Rehabilitation Act standards, guidance, best practices, tools, and checklists developed by the Access Board and the General Services Administration?
   - Resources from the World Wide Web Consortium (W3C), including tools for evaluating website accessibility?

XI. ENDNOTES

[1] The term "information and communication technology" (ICT) includes information technology and any equipment or interconnected system or subsystem of equipment, which is used in the creation, conversion, or duplication of data or information. ICT also includes information technology and any equipment or interconnected system or subsystem of equipment, which is used in the automatic acquisition, storage, analysis, evaluation, manipulation, management, movement, control, display, switching, interchange, transmission, reception or broadcast of data or information. The term includes, but is not limited to, electronic content, including email, electronic documents and Internet and Intranet websites; telecommunications products, including video communication terminals; computers and ancillary equipment, including external hard drives; software, including operating systems and applications; information kiosks and transaction machines; videos; IT services; and multifunction office machines that copy, scan, and fax documents. [36 CFR 1194.4]

[2] Statement of Samuel R. Bagenstos, Principal Deputy Assistant Attorney General for Civil Rights, Department of Justice before the Subcommittee on the Constitution, Civil Rights and Civil Liberties, Committee on the Judiciary, U.S. House of Representatives concerning *Emerging Technologies and the Rights of Individuals with Disabilities* (April 22, 2010). [Hereinafter referred to as DOJ Testimony] at page 1.

[3] DOJ Testimony at page 8.

[4] The three quotes included in the text may be found in the Preamble to the DOJ Advance Notice of Proposed Rulemaking regarding *Nondiscrimination on the Basis of Disability; Accessibility of Web Information and Services of State and Local Government Entities and Public Accommodations* 75 Federal Register 43460 (July 26, 2010). [Hereinafter referred to as DOJ Web Accessibility ANPRM]

[5] *See* for example *Accessibility: A Guide for Businesses and Organizations—Empowering employees, customers and partners with accessible technology.* Microsoft (2011) at pages 8-10. [Hereinafter referred to as the Microsoft Accessibility Guide]

[6] *Roadmaps for Enhancing Employment of Persons with Disabilities through Accessible Technology,* developed by participants at the Business Dialogue on Accessible Technology and Disability Employment (September 2007) at page 9-10. [Hereinafter referred to as *Roadmaps]*

[7] *Roadmaps* at page 10.

[8] Additional benefits of accessible design are discussed in *Developing a Web Accessibility Business Case for Your Organization* available at http://www.w3.org/WAI/bcase/.

[9] The definitions are taken from several sources, including *Electronic and Information Technology Accessibility Standards* (final rule codified at 36 CFR Part 1194) [hereinafter referred to as the Section 508 Accessibility Rule]; *Draft Information and Communication Technology (ICT) Standards and Guidelines* (published in the Federal Register on March 22, 2010 at 75 Federal Register 13457 [hereinafter referred to as the Section 508 ANPRM]; and the *BuyAccessible Wizard Glossary* https://app.buyaccessible.gov/DataCenter/Glossary.jsp.

[10] http://www.accessibletech.org/access_articles/general/whatIsAccessibleEIT.php.

[11] http://www.ada.gov/reg2.html.

[12] See http://www.access-board.gov/508.htm

[13] OFCCP Program Directive Number 281 (July 10, 2008) regarding Contractors' Online Application Selection Systems.

[14] *See* Preamble to DOJ Web Accessibility ANPRM at 75 Federal Register 43462. See also Web Accessibility Initiative of the World Wide Web Consortium, http://www.W3.org/WAI/intro/people-use-web/; Silverstein, Robert. *A Technical Assistance Guide for Federal Contractors Regarding Implementation of Online Application Systems that Meet the Needs of Qualified Individuals with Disabilities and Qualified Disabled Veterans,* Economic Systems, Inc (July 2009); and Microsoft *Accessibility Guide.*

[15] The summary description set out in the text regarding the duty to take affirmative action is derived from a review of multiple sources, including the regulations implementing Section 503 [41 CFR 60-741.43] and VEVRAA [41 CFR 60-300.43]; the OFCCP Federal Contract Compliance Manual; the Department of Labor, Office of Administrative Law Judges, Office of the Solicitor (Civil Rights Division) Index of Administrative Decisions Under Section 503:Topic 25—Affirmative Action; and EEOC Management Directive 715 (which clarifies affirmative action obligations of federal agencies under Section 501 of the Rehabilitation Act). *See Affirmative Action for People with Disabilities—Volume II Modernizing the Affirmative Action Provisions of the Section 503 and VEVRAA Regulations,* Economic Systems, Inc (April 30, 2010) at pages 49-50. [Hereinafter referred to as AAP Study Volume II]

[16] 42 U.S.C. 12101.

[17] 29 CFR Part 1630. In particular, see 29 CFR 1630.4, 1630.6 and 1630.9.

[18] 29 U.S.C.791.

[19] 29 U.S.C.793. *See* OFCCP Program Directive Number 281 (July 10, 2008) regarding Contractors' Online Application Selection Systems.

[20] 29 U.S.C.794

[21] *See* Preamble to DOJ Web Accessibility ANPRM, 75 Federal Register 43463.

[22] *See* Preamble to DOJ Web Accessibility ANPRM, 75 Federal Register 43464.

[23] 75 Federal Register 43460.

[24] 29 U.S.C.794d.

[25] 36 CFR Part 1194.

[26] 75 FR 13457.

[27] Telecommunications Act Accessibility Guidelines; Electronic and Information Technology Accessibility Standards, 76 Fed. Reg. 76,640 (proposed Dec. 8, 2011)(to be codified at 36 C.F.R. §§ 1193, 1194).

[28] This description of the WAI of the W3C to develop the WCAG is included in the Preamble to DOJ Web Accessibility ANPRM, 75 Federal Register 43463.

[29] Techniques for WCAG 2.0 http://www.w3.org/TR/WCAG20-TECHS/

[30] WCAG 2.0 At A Glance http://www.w3.org/WAI/WCAG20/glance/

[31] *See* DOJ Testimony at page 6 and Preamble to DOJ Web Accessibility ANPRM, 75 Federal Register 43464 and 43465.

[32] http://www.hhs.gov/ociio/regulations/joint_cms_ociio_guidance.pdf. *See* also reference to WCAG 2.0 standards in the recent Settlement Agreement between the National Federation of the Blind of California and the Law School Admission Council with the concurrence of the Department of Justice (April 25, 2011).

[33] *See* Action Step 7 for links to GSA's resources, including the BuyAccessible Wizard www.buyaccessible.gov.

[34] *See* Action Step 7 for links to W3C resources.

[35] *Dear Colleague Letter to elementary and secondary schools,* http://www2.ed.gov/about/offices/list/ocr/letters/colleague-201105-pse.html and *Frequently Asked Questions (FAQ),* http://www2.ed.gov/about/offices/list/ocr/docs/dcl-ebook-faq-201105.html.

[36] *Roadmaps* at page 9.

[37] The examples are taken from Microsoft's *Accessibility Guide, Accessible Information and Communication Technologies, Benefits to Business and Society* at pages 5-10, Roadmaps at pages 9-10, OneVoice for accessible ICT (2009-2010), and *Becoming an Employer of Choice—how accessibility can help you attract, recruit, and retain maturing workers,* IBM April 24, 2007. *See also* information on customizing a Web accessibility business case available in *Developing a Web Accessibility Business Case for Your Organization,* http://www.w3.org/WAI/bcase/ and EARN's *Talent to Drive Your Business's Success,* http://www.askearn.org/BusinessCase/

[38] The descriptions in the text are taken from the following document prepared by the Access Board: *Electronic and Information Technology Standards: An Overview,* www.access-board.gov/sec508/summary.htm

[39] *See* Access Board Advance Notice of Proposed Rulemaking updating and modernizing the Telecommunications Act Accessibility Guidelines and the Electronic and Information Technology Standards [75 *Federal Register* 13457 (March 22, 2010)]

[40] *See* 36 CFR 1194.21.

[41] DOJ Testimony at page 5.

[42] Judy Brewer, *Achieving the Promise of the Americans with Disabilities Act in the Digital Age—Current Issues, Challenges, and Opportunities,* Hearing before the Subcommittee on the Constitution, Civil Rights, and Civil Liberties, House Committee on the Judiciary, (April 22, 2010).

[43] 36 CFR 1194.22.

[44] *See* DOJ *Section 508 Survey, FY 2010,* U.S. Department of Justice Section 508 Survey Questions FY2010; *Recruitment Hints and Tips: Introduction; Recruitment Hints and Tips: Website access—Putting in place an action plan.* http://www.barrierfree-recruitment.com.

[45] *See* Action Step 1.

[46] 36 CFR Part 1194.

[47] http://www.w3.org

[48] See http://www.buyaccessible.gov; http://www.w3.org.

[49] 36 CFR 1194.23.

[50] http://www.accesswireless.org/Find/Gari.aspx.

[51] 36 CFR 1194.24.

[52] 36 CFR 1194.25.

[53] 36 CFR 1194.26.

[54] 36 CFR Subpart D.

[55] *See* Silverstein, Robert. *A Technical Assistance Guide for Federal Contractors Regarding Implementation of Online Application Systems that Meet the Needs of Qualified Individuals with Disabilities and Qualified Disabled Veterans,* Economic Systems, Inc (July 2009).

[56] *See* OFCCP Program Directive Number 281 (July 10, 2008) regarding Contractors' Online Application Selection Systems.

[57] *See* OFCCP Program Directive Number 281 (July 10, 2008) regarding Contractors' Online Application Selection Systems.

[58] *Recruitment Hints and Tips: Assisting people who cannot apply online—a manual fail-safe.* http://www.barrierfree-recruitment.com.

[59] *See* OFCCP Program Directive Number 281 (July 10, 2008) regarding Contractors' Online Application Selection Systems.

[60] *See* Appendix A to the Section 503 Regulation—*Guidelines on a Contractor's Duty to Provide Reasonable Accommodation.*

[61] *Roadmaps* at page 11.

[62] *Roadmaps* at page 9.

[63] *Roadmaps* at page 12.

[64] *Recruitment Hints and Tips: Minimizing the risk of outsourcing.* http://www.barrierfree-recruitment.com.

[65] *Guidance on Creating 508 Compliant IT Solicitations,* GSA's Section 508 Program. Guidance on Creating 508 Compliant IT Solicitations.

[66] The BuyAccessible Wizard http://www.buyaccessible.gov helps define the specific requirements of a particular ICT deliverable. It also provides a means to convey these requirements to the vendor. Once accessibility requirements are defined, the BuyAccessible Wizard also

provides the appropriate language to include these in the solicitation. The Wizard also provides generally accepted "good practices" to follow when assembling the solicitation, including:

Section 508 applicability statements, notifying when an exception is claimed, requesting accessibility information from vendors, and recommending an appropriate format for this information via a Voluntary Product Accessibility Template (VPAT) or a Government Product and Services Accessibility Template (GPAT). The Wizard also generates three additional guides to aid in the procurement of conformant EIT: Evaluation Guide (helps buyers evaluate various proposals based on commercial availability of their applicable provisions); Acceptance Guide (helps buyers evaluate various proposals based on commercial availability of their applicable provisions); and Design Guide (provides EIT developers with resources (internet links) that are qualified design and development methods to help ensure conformance to the applicable provisions based on generally accepted design and development methods).

[67] *See* http://www.section508.gov/index.cfm?fuseAction=BuyAccessiblePO.

[68] *Roadmaps* at page 12.

[69] *Roadmaps* at pages 11-12; Microsoft *Accessibility Guide* at page 42 and 50

[70] *Affirmative Action for People with Disabilities—Volume II Modernizing the Affirmative Action Provisions of the Section 503 and VEVRAA Regulations,* Economic Systems, Inc (April 30, 2010) at pages 49-50. [Hereinafter referred to as AAP Study Volume II]

[71] *Settlement Agreement Between the United States of America and Wells Fargo Company under the Americans with Disabilities Act,* DJ#202-11-239 (2011) at page 9.

[72] *Roadmaps* at pages 11-12.

[73] *Ensuring the Accessibility of Federal Electronic and Information Technologies Procured by Federal Agencies,* Memorandum for Chief Acquisition Officers and Chief Information Officers from Executive Office of the President (November 6, 2007).

[74] *Improving the Accessibility of Your Web Site, W3C Web Accessibility Initiative,* http://www.w3.org/WAI/impl/improving.html

[75] *Improving the Accessibility of Your Web Site, W3C Web Accessibility Initiative,* http://www.w3.org/WAI/impl/improving.html

[76] *Roadmaps* at page 12.

[77] *Affirmative Action for People with Disabilities—Volume II Modernizing the Affirmative Action Provisions of the Section 503 and VEVRAA Regulations,* Economic Systems, Inc (April 30, 2010) at pages 44-46. *See* also *DOJ Wells Fargo Settlement* at page 16.